T0207497

Wireless Sensor Networks

Rastko R. Selmic · Vir V. Phoha
Abdul Serwadda

Wireless Sensor Networks

Security, Coverage, and Localization

 Springer

Rastko R. Selmic
Louisiana Tech University
Ruston, LA
USA

Vir V. Phoha
Department of Electrical Engineering and
   Computer Science
Syracuse University
Syracuse, NY
USA

Abdul Serwadda
Department of Computer Science
Texas Tech University
Lubbock, TX
USA

ISBN 978-3-319-83581-5   ISBN 978-3-319-46769-6   (eBook)
DOI 10.1007/978-3-319-46769-6

Printed on acid-free paper

This Springer imprint is published by Springer Nature
The registered company is Springer International Publishing AG
The registered company address is: Gewerbestrasse 11, 6330 Cham, Switzerland

To Sue, Mila, Maksim, and Taylor.

To Li, Shiela, Rekha, Krishan, and Vivek.

Preface

Sensors represent a basic building block of technology systems we depend much on in our daily activities such as mobile phones, smart watches, smart cars, home appliances, etc. To date Wireless Sensor Networks (WSNs) represent perhaps the most widely deployed and highly explored networks that use sensors as part of their systems. It is through such a network that sensors communicate, share and fuse information, and thus provide foundations for applications such as large scale monitoring, surveillance, home automation, etc. With the advent of Internet of Things (IoT) and wearable devices embedded with sensors, new and exciting applications of WSNs have emerged. We expect a greater convergence of WSNs with these exciting new and emerging technologies such as the IoT.

Through this book, we not only present a structural treatment of the building blocks of WSNs, which include hardware and protocols architectures, but we also present systems-level view of how WSNs operate including security, coverage and connectivity, and localization and tracking. One can use these blocks to understand and build complex applications or pursue research in yet open research problems. The areas are wide: how one may deploy the wireless sensor nodes? How sensor nodes within the wireless network communicate with each other? What is their architecture? What are the security issues? And many more questions and their answers are provided for general engineering and science audience.

The purpose of writing this book is to give a systematic treatment of foundational principles of WSNs. We believe that this treatment provides tools to build or program specialized applications and conduct research in advanced topics of WSNs. Since each of us has academic experiences, we present the material from a pedagogical view with each chapter providing a list of references and a list of short questions and exercises. The field is growing at such rapid pace that it is impossible to cover all new developments; therefore, each chapter provides information with a balance towards pedagogy, research advances, and an enough introduction of important concepts, such that an interested reader should she be interested to explore further can then refer to cited papers in the references.

Our discussions in the book are motivated by demands of applications, thus most of the material, especially in the later chapters, has applications in areas where sensor networks may be used or deployed.

Any student with a university undergraduate education in mathematics, physics, computer science, or engineering will feel comfortable following the material. Readers primarily interested in qualitative concepts rather than the underlying mathematics or the programming of WSNs can skip the more mathematical parts without missing the core concepts.

The book can serve a basis for one-semester to two-semester course in WSNs. One can focus on WSN foundations or WSN security or coverage and control. We suggest the following:

- One-semester course with a focus on *coverage and control* of WSNs: Chaps. 1–3, 5, 6, and 8.
- One-semester course with a focus on *security* of WSNs: Chaps. 1, 2, 4, 7, and 8.
- One-semester course with a focus on *foundations* of WSNs: Chaps. 1–3, 5, and 7.
- One-semester course with a focus on WSN *hardware*: Chaps. 1–3, 5, and 8.

For a two-semester sequence, one can pick and choose the chapters. For example, one scenario may be as follows: follow first three chapters in the first semester supplemented by parts of chapters on security, coverage or control. A more applied course may include Chap. 8 in the first semester replacing fully or partially the content from security, coverage, and control. In the second semester, Chaps. 4–6 can be covered supplemented by course projects.

The book is organized as follows.

Chapter 1 provides foundations and gives a general description of WSNs, most common application where WSNs are used and common communication protocols that are basis for a WSN.

Chapter 2 covers background material needed to understand WSN topology, protocols, routing, coverage, etc. We include basic mathematical models that are used later in the book such as Voronoi diagrams and Delaunay triangulations. This chapter is recommended to be studied before coverage and connectivity or localization and tracking are covered.

Chapter 3 presents a WSN architecture including both hardware structure and functional details of all major components in the sensor node and a layered network architecture and description of various protocols. When we discuss hardware components, we present each building block of a sensor node and their important functional principles. For instance, we list important and common sensors that engineers and scientist might encounter when dealing with WSNs, and discuss their sensing principles. Similarly, when we discuss medium access protocols, we talk about common protocols that are currently in use.

Chapters 1–3 cover a basic background related to WSNs. Chapter 4 is a more focused material related to WSN security issues. Why are WSN predisposed to various security threats and what are the most important vulnerabilities? We cover basic attack and defense strategies that are applicable to a WSN. When discussing security, robustness of the network is closely related to sensor faults, proper sensor fault detection and mitigation. Malicious data on one sensor node can be interpreted and detected as a fault within the next hop in multi-hop network.

Chapter 5 presents coverage and connectivity, two related characteristics of the network and important quality of service measures. We discuss basic mathematical models for coverage and connectivity and then study more in-depth theoretical concepts related to coverage holes. This is also important from the security point of view where any coverage hole in the sensor network might represent a vulnerability point.

Chapter 6 covers another advanced topics—localization and tracking as well as important algorithms that are used today in such applications.

Chapter 7 provides a quality of service overview. Here we acknowledge that some quality of service measures are already covered in other chapters, such as coverage, and we discuss in more details only such measures that have not been covered previously.

Chapter 8 presents WSN platforms that are in use, some that have more of a historic value at the moment, and some that witnessed their own evolution into other closely related products.

We have tried to find a balance between simplicity, depth of treatment, and covering enough material without the risk of appearing superficial. We hope that we have succeeded in this endeavor.

A part of the research covered in the book was supported by the Air Force Research Laboratory (AFRL) at WPAFB, Sensors Directorate. We thank Todd Jenkins and Atindra Mitra (late) at AFRL. We acknowledge the help and support of Jinko Kanno in preparing material included in the Coverage and Connectivity chapter and Md Enam Karim in preparing material for the QoS chapter. We appreciate the help, support, and guidance of Jennifer Malat and Susan Lagerstrom-Fife from Springer in preparing this book.

Ruston, USA   Rastko R. Selmic
Syracuse, USA   Vir V. Phoha
Lubbock, USA   Abdul Serwadda

Contents

Chapter 1
Introduction

The development of microcontrollers, communication technology, microelectromechanical systems (MEMS), and nanotechnology allowed for research and development of new systems for sensing and communication called wireless sensor networks. Such networks are characterized as ad hoc (no previous setup or supporting infrastructure is required), utilize novel communication protocols, cooperatively monitor phenomena of interest, and communicate recorded data to the central processing station, usually called the base station. As the word *wireless* indicates, such networks of sensors communicate using wireless communication channels, allowing for easy deployment, control, maintenance, and possible sensor replacements.

Wireless sensors in networked systems are often called *nodes*, as they are built of many more components than just sensors. Sensor nodes are, from a hardware perspective, small form-factor embedded computers coupled with a variety of sensors that are chosen by the user depending on the targeted application. Sensor nodes usually have built-in microprocessors or microcontrollers, power supply in form of a battery, a memory, a radio, communication ports, interface circuits, and finally sensors for specific applications. They are complex embedded devices that combine from computer, communication, networking, and sensors technologies.

Being a network of small computer-like embedded devices, wireless sensor networks are significantly different from general computer-based data networks such as the Internet or Ethernet. Wireless sensor networks (WSNs) do not have topologies that are characteristic for Local Area Networks (LAN) such as bus, ring, or star. They are mostly ad hoc networks deployed randomly in the field relying mostly on the widely adopted underlying IEEE 802.15.4 standard for embedded devices. They are application-specific with communication and networking sometimes specifically designed to accommodate targeted applications. Bounded by numerous constraints, usually not seen in general data networks, such as limited energy and bandwidth availability, small form-factor, large number of nodes deployed over wide open areas, and others, WSNs' networking and communication must be creatively adjusted to support specific applications which we discuss in

© Springer International Publishing AG 2016
R.R. Selmic et al., *Wireless Sensor Networks*,
DOI 10.1007/978-3-319-46769-6_1

later chapters. Thus, a new cross-layer optimization and changes in communication protocols have been developed to address specific requirements for sensor networks.

Exposed to numerous constraints, environmental and technological difficulties, and driven by market needs, WSNs have evolved and developed numerous characteristics that distinguish them from standard computer-based networked systems. They are capable of unattended operation with very limited or no supervision. The main sensor network components, the sensor nodes, are inexpensive and usually disposable. The sensor network supports dynamic topologies that can overcome node or sensor failures, drops in communication links, or movement of nodes. Nodes can also operate in harsh and dangerous environments with a human operator standing at a safe distance. Due to their small size and lack of cables, WSNs are not disruptive for the environment or industrial processes. Compared to individual sensors assigned to measure and observe specific phenomena of interest, sensor networks are capable of cooperative measurements and cooperative in-network data processing.

In the following sections, we first give an overview of the sensor networks which are a super set of the wireless sensor networks and then give brief details of wireless sensor networks and the applications of wireless sensor networks.

1.1   Sensor Networks

Sensor networks are composed of a large number of sensor nodes that are deployed to collectively monitor and report any phenomena of interest. In a sensor network, the physical layer specifies electrical and mechanical interface to the transmission medium and can be wired, wireless, or a combination of both. Sensor networks are a superset of WSNs and, as such, share some common attributes that are integral to all sensor network systems.

We first discuss general attributes of sensor networks and in subsequent sections focus on sensor networks with wireless signal transmission.

- **Phenomena of Interest:** Based on the domain or environment in which a sensor network operates, phenomena of interest can be purely physical (for example, leakage of hazardous plumes in a chemical factory, radiation activity leakage in a nuclear waste storage facility, occurrence of forest fires, etc.) or can be observable manifestations of a dynamical physical phenomenon (for example, occurrence of anomalies in aerial imagery due to aircraft jitter, occurrence of a runtime faults in an embedded system due to an ill-conceived electronic circuitry, etc.).

- **Composition and Type:** A sensor network can be homogenous (i.e., composed of same type of sensors) or heterogeneous (i.e., different types of sensors) in composition. Further, a sensor network can be a passive network, comprising sensors that detect phenomena via radiations emitted by an object or its surrounding environment (e.g., acoustic, seismic, video, and magnetic sensor networks) or an active sensor network comprising sensors that probe into the environment by sending signals and measuring responses (e.g., radar and lidar). A sensor network can be stationary (e.g., seismic sensor network) or mobile (e.g., sensors mounted on mobile robots and unmanned aerial vehicles).

- **Sensor Deployment:** It involves placing sensor nodes within the permissible neighborhood of the phenomena of interest, so that all defined constraints on the quality of sensing are satisfied. Based on the sensing environment, sensor network deployment can be planned (e.g., as in inventory storage facilities, nuclear power plants, etc.,) or ad hoc (e.g., air-dropped for monitoring movement in hostile territories).

- **Monitoring, Processing, and Reporting:** It involves communication and processing within groups of sensor nodes, base stations, command and control units, and all other entities that gather pertinent measurements about the phenomena of interest and eventually make decisions to actuate appropriate response. Communication can be wired or wireless, depending upon the application requirement and sensing environment. Similarly, depending on the target application, processing can be centralized, i.e., all data are sent to and processed by a centralized base station or autonomous, i.e., each node takes its own decision, or a hybrid, i.e., semi-autonomous or loosely centralized.

Fundamental advances in microelectromechanical systems (MEMS), fabrication technologies, wireless communication technologies, low-power processing, and distributed computational intelligence have led to the development of low-cost high-density sensor networks, which not only provide large spatial coverage and high-sensing resolutions but also have high levels of fault tolerance, endurance, and flexibility in handling operational uncertainty. Consequently, sensor networks are becoming ubiquitous in many application areas as diverse as military, health, environment and habitat monitoring, and home, to name a few. Below is a partial list of some application areas in which sensor networks have shown promising utility.

- **Military Applications:** Sensor network research was initially motivated by military applications such as monitoring equipment and inventory, battlefield surveillance and reconnaissance, target tracking, battlefield damage assessment, nuclear, chemical, and biological weapon detection and tracking, etc. Military applications demand rapid deployment, robust sensing in hostile terrains, high levels of longevity, energy conservation, and information processing to extract useful, reliable, and timely information from the deployed sensor network.

- **Environment Monitoring Applications:** Include chemical or biological detection, large scale monitoring and exploration of land and water masses, flood detection, monitoring air, land, and water pollution, etc.

- **Habitat Monitoring Applications:** Include forest fire detection, species population measurement, species movement tracking in biological ecosystems, tracking bird migrations, vegetation detection, soil erosion detection, etc.

- **Health Applications:** Include real-time monitoring of human physiology, monitoring patients and doctors in hospitals, monitoring drug administration, blood glucose level monitors, organ monitors, cancer detectors, etc.
- **Infrastructure Protection Applications:** Include monitoring nation's critical infrastructure and facilities (e.g., power plants, communication grids, bridges, office buildings, museums, etc.) from naturally occurring and human-caused catastrophes. Sensor networks in these applications are expected not only to provide reliable measurements to facilitate early detection but are also required to provide effective spatial information for localization.
- **Home Applications:** Sensor networks are being deployed in homes to create smart homes and improve the quality of life of its inhabitants. Recently, a new paradigm of computing, called 'ambient intelligence' has emerged with a goal to leverage sensor networks and computational intelligence to recreate safe, secure, and intelligent living spaces for humans.

Next, we briefly discuss some important design factors that arise in the application of sensor networks.

- **Fidelity and Scalability:** Depending on the operational environment and the phenomenon being observed, fidelity can encompass a multitude of quality or performance parameters such as spatial and temporal resolution, consistency in data transmission, misidentification probability, event detection accuracy, latency of event detection, and other quality of service-related measures. Scalability broadly refers to how well all the operational specifications of a sensor network are satisfied with a desired fidelity, as the number of nodes grows without bound. Depending on the measure of fidelity, scalability can be formulated in terms of reliability, network capacity, energy consumption, resource exhaustion, or any other operational parameter as the number of nodes increases. While it is very difficult to simultaneously maintain high levels scalability and fidelity, tuning sensor networks to appropriately tradeoff scalability and fidelity has worked well for most applications.
- **Energy Consumption:** Individual sensor nodes, electronic circuitry supporting the nodes, microprocessors, and onboard communication circuitry are the primary consumers of energy. In case of WSNs, the most likely energy source is a lithium-ion battery. Depending on the operational environment, energy constraints can be an important factor in sensor network design. In structured and friendly environments (e.g., industrial infrastructure, hospitals, and homes), specific arrangements can be conceived to replenish onboard batteries on individual nodes for WSNs. However, in harsh environments and large territories (typical in military and habitat monitoring applications), replenishing energy may be impractical or even impossible. In such situations, energy conservation becomes a critical issue for extending a sensor network's longevity. Energy conservation can be addressed at multiple levels, starting from the designing energy-aware sensors, energy-aware electronic circuitry to energy conserving communication, processing, and tasking.

- **Deployment, Topology, and Coverage:** Depending on the operational environment, constituent nodes in a sensor network can be deployed in a planned fashion (choosing specific positions for each node) or in a random fashion (dropping nodes from an aircraft). Deployment can be an iterative process, i.e., sensors can be periodically added into the environment or can be a one-time activity. Deployment affects important parameters such as node density, coverage, sensing resolution, reliability, task allocations, and communications. Based on the deployment mechanism, environment characteristics, and operational dynamics, a sensor network's topology can range from static and properly defined to dynamic and ad hoc. In some environments, the topology of a sensor network can be viewed as a continuous time dynamical system that evolves (or degrades) over time largely due to exogenous stimuli or internal activity (for example, node tampering is an external stimuli while power exhaustion is an internal activity—both have a potential to drastically change the topology of the sensor network). In its simplest form, a sensor network can form a single-hop network with every node communicating with its base station. Centralized sensor networks of this kind form a star-like network topology. A sensor network may also form an arbitrary multi-hop network, which takes two or more hops to convey information from a source to a destination. Multi-hop networks are more common in mobile sensing, where the ad hoc topology demands message delivery over multiple hops. Topology affects many network characteristics such as latency, robustness, and capacity. The complexity of data routing and processing also depends on the topology. Coverage measures the degree of coverage area of a sensor network. Coverage can be sparse, i.e., only parts of environment fall under the sensing envelope, or dense, i.e., most parts of environment are covered. Coverage can also be redundant, i.e., the same physical space is covered by multiple sensors. Coverage is mainly determined by the sensing resolution demands of an application.

- **Communication and Routing:** Because sensor networks deal with limited bandwidth, processing, and energy, operate in highly uncertain and hostile environments (e.g., battlefields), constantly change topology and coverage, lack global addressing, and have nodes that are noisy and failure-prone, traditional Internet communication protocols such as Internet Protocols (IP), including mobile IP may not be adequate. Most communication specifications originate from answering the following question: Given a sensor network, what is the optimal way to route messages so that the delivery between source and destination occurs with a certain degree of fidelity? Many routing schemes have been proposed, with each routing scheme optimizing a suitable fidelity metric (e.g., sensing resolution) under defined constraints of operation (e.g., energy constraints). Popular routing schemes include data-centric routing, in which data are requested on demand through queries to specific sensing regions (e.g., directed diffusion, SPIN: Sensor Protocols for Information via Negotiation, CADR: Constrained Anisotropic Diffusion Routing, and ACQUIRE: Active Query Forwarding In Sensor Networks); flooding, and gossiping, which are based on broadcasting messages to all or selected neighbor nodes; energy-aware routing (e.g., SMECN: Small Minimum Energy Communication Network, GAF: Geographic Adaptive Fidelity, and GEAR: Geographic and Energy Aware Routing); hierarchical routing, in which messages are passed via multi-hop communication within a particular cluster and by performing data aggregation and fusion to decrease the number of transmitted messages (e.g., LEACH: Low-Energy Adaptive Clustering Hierarchy, PEGASIS: Power-Efficient Gathering in Sensor Information Systems, and TEEN: Threshold Sensitive Energy Efficient Sensor Network Protocol).

- **Security:** Security Requirements of a sensor network encompass the typical requirements of a computer network plus the unique requirements specific to the sensor network application. Security in sensor networks aims to ensure data confidentiality—an adversary should not be able to steal and interpret data/communication; data integrity—an adversary should not be able to manipulate or damage data; and data availability—an adversary should not be able to disrupt data flow from source to sink. Sensor networks are vulnerable to several key attacks. Most popular are eavesdropping (adversary listening to data and communication), denial-of-service attacks (range from jamming sensor communication channels to more sophisticated exploits of 802.11 MAC protocol), Sybil attack (in which malicious nodes assume multiple identities to degrade or disrupt routing, data aggregation, and resource allocation), traffic analysis attacks (aim to identify base stations and hubs within a sensor network or aim to reconstruct topologies by measuring the traffic flow rates), node replication attacks (involves adding a new node which carries the id of an existing node in the sensor network to mainly disrupt routing and aggregation), and physical attacks (range from node tampering to irreversible node destruction). Several defenses have been proposed against attacks on sensor networks. Solutions that ensure data confidentiality use energy-aware cryptographic protocols, which are mostly based on Triple-DES, RC5, RSA, and AES algorithms. Defenses against denial-of-service attacks include rouge node identification and elimination, multi-path routing, and redundant aggregation. Primary defenses against Sybil attacks are direct and indirect node validation mechanisms. In direct validation a trusted node directly tests the joining node's identity. In indirect validation, another two levels of trusted nodes are allowed to testify for (or against) the validity of a joining node. Defenses against node replication attacks include authentication mechanisms and multicast strategies, in which new nodes are either authenticated through the base station or (in the case of multicast strategy) the new nodes are authenticated via a group of designated nodes called 'witnesses'. Strategies to combat traffic analysis attacks include random walk forwarding, which involves occasionally transferring messages to a pseudo base station, fake packet generation, and fake flow generation.

1.2   Wireless Sensor Networks

1.2.1   Historical Perspective, Aloha Networks

The first experiment with wireless signal transmission was carried out in 1893 by Nikola Tesla. A few years later, Tesla was able to remotely control small boats, setting a path for the later development of guided missiles and precision-guided weapons. The first amplitude modulation (AM) signals were generated in 1906 and high frequency radio signals in 1921. Armstrong is credited for development of first frequency modulated signal 1931. Metcalfe and Boggs at Xerox PARC are credited for creation of Ethernet in 1973 with an initial transmission rate of 2.9 Mbit/s. That was later a foundation for creation of IEEE 802.3 standard that is still being developed and expanded. In 1997, IEEE 802.11 standard was created with a bandwidth of 2 Mbit/s with subsequent modification and addition to the standard. In 1999, 802.15.1, commonly called Bluetooth, was formulated for short-range wireless communication between embedded devices.

Aloha communication scheme, invented by Norman Abramson in 1970 at the University of Hawaii [1], was one of the first networking protocols that successfully networked computer systems, in this case different campuses of the University of Hawaii on different islands. The concepts are widely used today in Ethernet and sensor networks communications, Figs. 1.1 and 1.2. The protocol allows computers on each island to transmit a data packet whenever there is a packet ready to be sent. If the packet is received correctly, the central computer station sends an acknowledgment. If the transmitting computer does not receive the acknowledgment after some time due to transmission error, which can be due to collision of packets transmitted at the same time from another system, the transmitting computer resends the packet. This process is repeated until the sending computer receives the acknowledgment from the central computer. The protocol works well for simple networks with low number of transmitting stations. However, for networks with multiple nodes, the protocol causes small throughput due to increase of collisions.

Fig. 1.1 Pure Aloha protocol where nodes transmit packets randomly with possible collisions with packets from other nodes (*gray*)

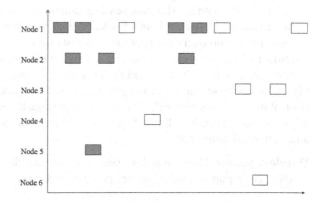

Fig. 1.2 Slotted Aloha
protocol where nodes transmit
packets only at pre-assigned
time intervals; however, the
collisions with packets from
other nodes are still possible
(*gray*)

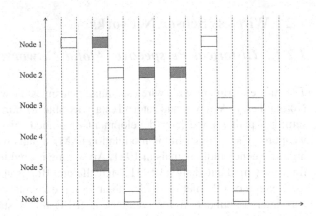

A modification of the algorithm allows nodes to transmit the same size packets only at pre-specified slot boundary. In this case transmission is not completely random and the number of collisions is reduced in half compared to the pure aloha protocol.

Aloha protocols fall into the category of contention-based protocols where there is a possible contention between nodes on the network (all nodes contend for the channel causing possible collisions). Other Medium Access Control (MAC) contention-based protocols include multiple access collision avoidance protocol (MACA), modified version such as multiple access with collision avoidance for wireless (MACAW), busy tone multiple access (BTMA), floor acquisition multiple access (FAMA), IEEE 802.11, and others [22].

1.2.2 Background on Wireless Sensor Networks

Wireless sensor networks (WSNs) are networks of autonomous sensor devices where communication is carried out through wireless channels. WSNs have integrated computing, storing, networking, sensing, and actuating capabilities [2, 3, 6, 7, 8, 20, 30, 35] with overlapping sensing, computing, and networking technologies (other important references and books in this area are given at the end of the section; for a good overview paper see for instance [19]). These networks consist of a number of sensor nodes (static and mobile) with multiple sensors per node that communicate with each other and the base station through wireless radio links (see Fig. 1.3). The base station, or the gateway, is used for data processing, storage, and control of the sensor network. Sensor nodes are usually battery powered; hence the whole sensor network is limited by fundamental tradeoffs between sampling rates and battery lifetimes [20].

Wireless Sensor Node Wireless sensor nodes are the main building blocks of WSNs. Their purpose is to "sense, process, and report". Requirements for sensor

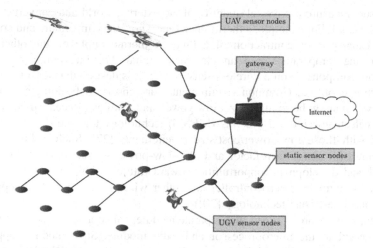

Fig. 1.3 Ad hoc wireless sensor network with static and mobile nodes placed on unmanned ground vehicles (UGVs) or unmanned aerial vehicles (UAVs)

nodes are to be small, energy efficient, and capable of in situ reprogramming. Examples include sensor nodes such as MICA motes from MEMSIC, Moteiv from Sentilla, EmbedSense from MicroStrain, Inc., and others. Figure 1.4 shows two typical sensor nodes.

Sensor nodes consist of a variety of sensors (sometimes built in on a separate module called sensor module), a microcontroller for on-board communication and signal processing, memory, radio transceiver with antenna for communication with neighboring nodes, power supply, and supporting circuitry and devices. Most of the sensor nodes run their own operating system developed for small form-factor, low-power embedded devices, such as TinyOS [5] or embedded Linux, that provides inter-processor communication with the radio and other components in the system, controls power consumption, controls attached sensor devices, and provides support for network messaging and other protocol functions.

Fig. 1.4 Sensor node MICA2 (*left*, *source* MEMSIC) and Tmote Sky (*right*, *source* www. advanticsys.com)

Sensors measure a physical quantity of the external world and convert it into a readable signal. For example, a thermometer measures the temperature and converts it into expansion or contraction of a fluid. A thermocouple on the other hand converts the temperature into an electronic signal. For integration with other electronic components on a wireless sensor node, it is desired that sensors produce an electronic output. Common requirements for sensors to be integrated with a WSN system are to be small in size, low power, and low cost. Recent advancements in microelectromechanical systems (MEMS) technology allowed development of sensors with those low-power/cost/size requirements [27]. Such technology not only allows for small form-factor and ultra-low-power sensors devices, but opens research and development opportunities toward future on-chip integration of sensors, radio, memory, microcontroller, and other wireless sensor node components (see about Smart Dust technology [30]).

Often, sensors are grouped in a separate module, called a sensor board, that can be connected to the microprocessor and radio module. Such modular approach allows users to combine different sensors with the same WSN platform, thus minimizing the development time for new applications, for instance, Fig. 1.5 shows Louisiana Tech University sensor board connected to MEMSIC Technology's Mica2 radio module. The sensor module supports three chemical sensors that can detect three chemical agents simultaneously, namely CO, NO_2, and CH_4.

Gateway/Base Station The gateway or the base station for wireless communication provides sensor data collection into a database. The radio transceiver of the base station is communicating with the sensor nodes in the field. The base station is a stand-alone system with a chassis that is against an inhospitable environment. The gateway/base station provides *gateway* connection with other networks. If possible, base station is connected to the Internet, thus allowing some remote system monitoring and data acquisition. It can run database software applications for the management of sensing data. Base station sub-system can host any user interface application accessible through Internet or locally at the base station.

Wireless Sensor Network Protocols Wireless sensor network protocols are designed to accommodate specific features and properties of wireless sensor networks including their geographically distributed deployment, self-configuration,

Fig. 1.5 Louisiana Tech
Univ. wireless sensor node for
chemical agent monitoring
applications built on
MEMSIC Mica2 platform

energy constraints usually limited by battery supply, wireless communication in often noisy environment, long lifetime requirements, and high fault tolerance. Most of the protocols are specific for or related to one of the features of sensor networks. An overview of wireless sensor networks protocols is provided in [23].

Medium Access Control (MAC) Initial MAC protocols such as Aloha [39] stem from computer network protocols. Such protocol for wireless sensor networks is given in [7]. The drawback of such protocols is that the on-board processor consumes power during idle periods. A suggested improvement is to avoid listening to the channel when it is idle. This could be implemented by transmitting signals having a preamble in front of sent packets. On waking up periodically to check the signal preamble, the receiver decides if it needs to be active or can continue to sleep.

Other examples of MAC protocols include Carrier Sense Multiple Access (CSMA), where a transmitter listens for a carrier signal before trying to send packets. In this scheme, the transmitter tries to detect or "sense" a carrier before attempting to transmit. If there is a carrier in a medium, the node wishing to transmit waits for the completion of the present transmission before initiating its own transmission. Sensor-MAC (S-MAC) [33, 34] is a protocol designed for wireless sensor networks that supports energy conservation of nodes and self-configuration, and its variations such as Timeout-MAC (T-MAC), DMAC, TRaffic-Adaptive Medium Access (TRAMA), and others [23]. In S-MAC protocol all nodes go to a sleep mode periodically. If a node wants to communicate with its neighbor, it must contend with other neighbors of the destination node for the communication medium. The transmitting node waits for the destination node to wake up, and sends Request to Send (RTS) packet. If the packet is received successfully, node wins the medium, and receives Clear to Send (CTS) packet. Each node maintains a sleep schedule for its neighbors through synchronization process, carried out by periodically sending a synchronization packet. The duty cycle of sleep schedule is fixed. Improvements of S-MAC such as Pattern-MAC [36] offer adaptable sleep–wake up schedule for sensor nodes.

Standard medium access control protocols include Time Division Multiple Access (TDMA) and Frequency Division Multiple Access (FDMA) [24, 26]. In TDMA, radio transmits in specifically allocated time slots. Duty cycle of the radio is reduced, and energy efficiency improved, since sensor nodes do not need to listen during idle periods. Microcontroller and transceivers can be in the sleep mode. TDMA protocol has some disadvantages when applied to ad hoc sensor networks. When the number of nodes changes, it is difficult for TDMA protocol to dynamically specify new time slots for new nodes. To alleviate the problem, a modified TDMA protocol [28] uses super frames where a node schedules different time slots to communicate with neighboring nodes. The problem with this communication scheme is a low bandwidth where the node cannot reuse time slots allocated for communication with some other sensor node.

In terms of routing protocols, a shortest radio path algorithm was proposed in [32] where the metric used is the received signal strength. Each radio receiver has

the coded information about the strength of the signal, enabling the receiver to find
the closest sensor node in the field and communicate with it. The base station starts
initialization process, where all sensor nodes identify themselves and thus identify
distance between each other. This way all sensors can be located with respect to the
base station.

1.3   WSN Applications

Convenience and cost savings of wireless communication, the small form factor of
microprocessors, microcontrollers, memory, radio, and other electronic components, and variety of sensors developed recently as a result of advancement in
MEMS and other sensor technologies, allowed for a broad adoption of WSNs in a
range of applications in many different areas. Here, we list a few examples of
deployed WSNs from different application domains.

Wireless Sensor Networks for Habitat Monitoring Deployed for habitat monitoring on Great Duck Island of the coast of Maine [20], this sensor network testbed
was one of the first applications of WSNs used in real time in the wild. A team from
Intel Corporation and University of California, Berkeley deployed 32 wireless
sensor nodes on Duck Island where the system was used for seabird colonies
monitoring. The advantage of this system is that it does not disrupt nature and
species being monitored.

The system has a hierarchical structure and wireless sensor nodes are deployed
in clusters or patches. Every cluster has a gateway, which transmits the data to one
central location, the base station, located on the island. Sensor nodes communicate
using multi-hop protocol where information hops from lower level leaves toward
the gateway. The base station has Internet connection through satellite two-way
communication link as well as database management for data processing and
storage. The architecture of the system for habitat monitoring is shown in Fig. 1.6.

Mica sensor nodes were used as the sensor network platform. Nodes are
equipped with 916 MHz radio, small form-factor Atmel ATmega103 [40]—an 8-bit
microcontroller that runs at 4 MHz, has 128 Kbytes of flash memory, and built-in
10-bit analog-to-digital converters, two batteries and other supporting circuitry.

The system can operate at least 9 months from non-rechargeable batteries.
Increased battery lifetime can be achieved using innovative methods for energy
harvesting from the environment (see for instance [12, 14, 25]), by applying
intelligent/adaptive control methods [29], and/or efficient coordination methods [4].
Sensor nodes can be reprogrammed in the field online, in situ. Sensor nodes are
equipped with light, temperature, infrared, humidity, and barometric pressure
sensors. Packaging that consists of acrylic enclosure has been developed specifically for this application. Proposed scheduled communication between sensor nodes
are the following:

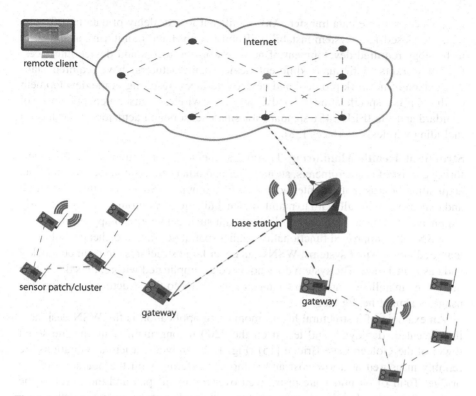

Fig. 1.6 Wireless sensor network monitoring system for habitat monitoring [20]

1. Nodes determine the number of hops (hop-level) from the gateway. Leaf nodes transmit first to the next level that has one less hop-level. After transmission is completed, sensor nodes go to a sleep mode where unused node components are shut down. The nodes are awaken again at a specific time instant, resembling to TDMA policy.
2. Nodes are awaken from the leaves toward the base station, independently of the nodes at the same hop-level. Data are passed from the leaves to the upper nodes in the network tree. The drawback is that the number of sub-trees and paths can be much larger than the number of hop-levels.
3. Low-power MAC protocols such as S-MAC [33, 34] and Aloha with preamble sampling [7] can also be used. They do not require communication scheduling but require additional energy and bandwidth for collision avoidance.

Industrial Control and Monitoring Compared with standard data networks where bandwidth, and therefore the data rate, is the most important network parameter, in industrial control and monitoring applications reliability and scalability are the most important performance measures. Monitoring and controlling temperature in an industrial boiler system does not require large data rate transfer; it necessitates reliable data transfer. Any significant loss or delay of data transfer can result in closed-loop system instability. Robust control and monitoring using WSN technology required development of new network protocols and device interfaces. Global markets, with many different device manufacturers, have required standardization in network protocol and device interfaces, resulting in the development of the ZigBee specification for IEEE 802.15.4 wireless sensor network protocol standard and the IEEE 1451 standard for smart sensors and actuators (transducers) including wireless interfaces [38].

Structural Health Monitoring Traditional methods for structural health monitoring consist of accelerometers, strain gages and other sensors connected to the data acquisition boards that are interfaced to a PC computer. Such systems are difficult and expensive to install, hard to maintain, and bulky to carry around. It is particularly expensive to achieve high spatial density with such conventional approach.

WSNs offer improved functionality, higher spatial density, and cheaper solutions than traditional wired systems. WSNs can cover large structures, and can be quickly and easily installed. The system does not need a complicated wiring, thus disruption due to the installation and maintenance of the WSN to the structure operation and usage is almost negligible.

An example of a structural health monitoring application is the WSN designed, implemented, deployed, and tested on the 4200 ft long main span and the south tower of the Golden Gate Bridge [15] (Fig. 1.7). Ambient structural vibrations are reliably measured at a low cost and without interfering with the operation of the bridge. Total of 64 nodes are distributed over the main span and the tower of the bridge. Sensor nodes measure vibrations with 1 kHz sampling rate, which was considered more than enough for civil structure monitoring applications. The accelerometer data are passed through low-pass anti-aliasing filter, fed into the analog-to-digital converter on the sensor node, and processed and transmitted wirelessly. The data are transmitted over a 56-hop network toward the base station. The system uses MicaZ sensor nodes with accelerometer sensor boards designed for this specific application that monitors acceleration in two directions. The nodes were packaged into plastic enclosing to protect it from gusty wind, fog, and rain, and installed on the bridge. Data sampling duty cycle is an order of magnitude

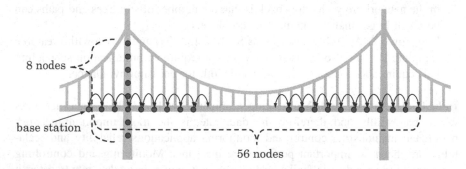

Fig. 1.7 Wireless sensor network used for structural monitoring at the Golden Gate Bridge [15]

higher than in environmental monitoring applications. Time synchronization across the network is required to correlate vibration measurements at different bridge locations. For larger network this can be challenging problem due to drift of clocks at each sensor node. The Flooding Time Synchronization Protocol [21] has been implemented to guarantee precise and coordinated measurements across the network. Embedded software is based on TinyOS operating system with newly developed software components.

Chemical Agents Monitoring Monitoring of chemical agents, their detection, and identification are of great importance for national security, homeland defense, consumer industry, and environmental protection. Being aware of potentially dangerous chemical agents in our surroundings can save our lives and provide crucial information for countermeasures. One of the challenges of emergency responses to weapons of mass destruction is to develop portable distributed sensor network capable of monitoring, detecting, and identifying different chemical agents at the same time. Important wireless sensor network requirements are multiple chemical agent detection and identification, distributed sensor network infrastructure, lightweight, and user-friendly.

The chemical agent monitoring applications are closely related to microelectromechanical systems (MEMS) technology that allows for small-form factor sensor arrays that can be easily integrated into low-power wireless sensor nodes, [11]. An example of MEMS chemical sensor that is suitable for WSN application is a microcantilever sensor using adsorption-induced surface tension that can be used to detect part-per-trillion (ppt) level of species both in air and solution. An electron micrograph of a cantilever and its structure are shown in Fig. 1.8 [11, 37].

The technology is based upon changes in the deflection and resonance properties induced by environmental factors in the medium in which a microcantilever is immersed. By monitoring changes in the bending and resonance response of the cantilever, mass and stress changes induced by chemicals can be precisely and accurately recorded. Usually MEMS sensors provide low-voltage signals, and interface electronics between chemical sensors and wireless sensor node is needed that includes signal conditioner (amplifier and filter) and signal multiplexer, Fig. 1.9.

ZnO

Silicon Oxide

Platinum

20KV 623X 16.1μ 3107

Fig. 1.8 Electron micrograph of microcantilever with a length of 200 μm (*left*) and structure of the microcantilever sensor (*right*)

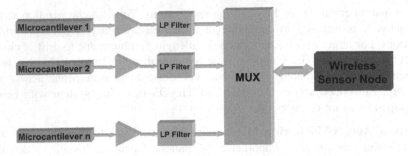

Fig. 1.9 Interface electronics for chemical sensor array on wireless sensor node

Military Applications Due to its small form-factor, possibility of ad hoc deployment, and no strict requirement for other power or communication infrastructure, WSNs find numerous use in military applications. For instance, shooter localization in urban environment cannot be accurately estimated with standard, centralized-based approach due to large multipath effects and limited coverage area. WSNs provide technology platform for a distributed solution where acoustic sensor data are cooperatively processed to estimate the shooter localization in an urban environment [16, 17].

The system [17] consists of a WSN with acoustic sensors, implemented as a custom-based sensor boards with DSP or FPGA devices, measures shockwaves and their time of arrivals. The measured data are sent to the base station for data fusion and shot trajectory estimation based on collected information from distributed sensor network. Time synchronization among sensor nodes and their known deployment location allows for accurate fusion of acoustic measurements and localization of the shooter or multiple shooters. The system can easily be extended into self-localizing sensor network where sensor nodes will localize themselves in real time using GPS or other localization techniques and then use sensor data to estimate the shooter location.

Surveillance Applications Such applications leverage recent technology advancements in WSNs to effectively and safely study volcanic activities [31]. An example of such a system is deployed to monitor Tungurahua volcano in central Ecuador. Scientists collect seismic data to monitor and study volcanic activity. To distinguish the volcano eruption with earthquakes or mining explosions, a correlation of infrasonic and seismic events is needed. Wireless sensor nodes can be placed close to the volcano crater and transmit the data to the base station on a safe distance for future processing.

The WSN consists of sensor nodes equipped with a specially constructed microphone to monitor infrasonic (low-frequency acoustic) signals from the volcanic vent during eruptions. Data are transferred to a gateway that forwards data wirelessly using long-range radio link to the base station at the volcano observatory. Time synchronization is achieved using a GPS node that supplies other nodes and the gateway with the timestamp data, Fig. 1.10.

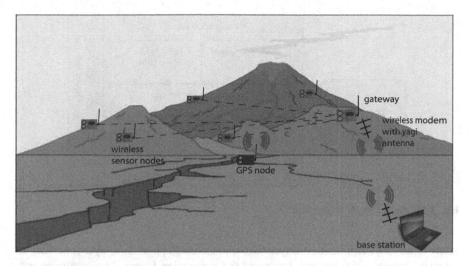

Fig. 1.10 WSN used for monitoring volcano activities [31]

Since volcano data are sampled at a higher rate than environmental sensor network applications (100 Hz), in-network data aggregation and distributed event detection is required. Such constraints require precise time synchronization, either using extra GPS equipped node or time-synchronizing protocols, and correlation of data among spatially close sensor nodes. Sensor nodes communicate with their neighbors to determine if an event of interest has occurred, [31] based on a decentralized voting scheme. Nodes keep track of window of data and also run event detection algorithm. In case a local event occurs, the node broadcasts a vote. If a node receives sufficient number of votes, a global data collection starts. This distributed event detection reduces the bandwidth usage and allows larger spatial resolution and larger sensing coverage areas. Sensor nodes are enclosed in water-proof packaging with antennas sealed with silicone.

1.4 WSN Common Communication Standards

ZigBee is a standard developed for low-power WSN monitoring and control applications which require reliable and secure wireless data transfers. It uses the existing IEEE 802.15.4 Physical layer and Medium Access Control sub-layer while adding networking, routing, and security of data transfers. It supports multi-hop routing protocols that can extend the network coverage. The physical layer operates at 868 MHz, 20 Kbps (Europe), or 915 MHz 40 Kbps (USA), and 2.4 GHz, 250 Kbps. Direct sequence spread spectrum is used with offset-quadrature phase shift keying modulation at 2.4 GHz band or with binary-phase-shift keying modulation at 868 and 915 MHz bands. Figure 1.11 shows ZigBee layered stack

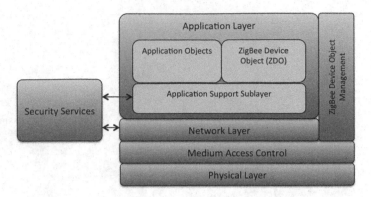

Fig. 1.11 ZigBee stack architecture [39]

architecture. The Application Layer has Application Support Sub-layer (provides an interface between the network layer and the application layer), Application objects (defined by manufacturers), and ZigBee Device Object (an interface between the application objects, the device profile, and the application support sub-layer, responsible for initialization of application support sub-layer, the network layer, and security services as well as processing configuration information from applications).

ZigBee's network layer allows for mesh, star, and tree topologies. The mesh topology supports peer-to-peer communication. In a star topology, there is a network Coordinator node that initiates and maintains devices on the network and can connect to other networks, [39]. In tree topology, Router Devices are responsible for moving data and control messages. ZigBee End Device communicates with the coordinator or router and cannot be used for hopping data from other devices.

ZigBee offers improved security features over IEEE 802.15.4 protocol—it uses a 128-bit Advanced Encryption Standard-based algorithm. It provides mechanisms for moving security keys around the network, key establishment, key transport, frame protection, and device management. These services form the building blocks for implementing security policies within a ZigBee device.

The IEEE 1451 standard for smart sensors and actuators was developed under leadership from the National Institute of Standards and Testing (NIST). A detailed description of the standards is given in [13, 18]. This standard is also used in integrated system health management [9] and in smart actuator control with transducer health monitoring capabilities [10]. The standard has been divided into six subgroups. IEEE 1451.0 defines a set of commands, operations, and transducers electronic data sheets for the overall standard. The access to the devices is specified and it is independent of the physical layer. IEEE 1451.1 defines communication with the Network Capable Application Processor (NCAP). This part of the standard specifies client–server or client–client type of communication between NCAP and other network devices, or between several NCAPs as is often case in a complex system with many smart sensors and actuators. IEEE 1451.2 includes the definition of Transducer Electronic Data Sheets (TEDS) and an interface between transducer

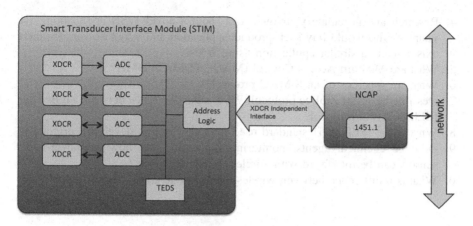

Fig. 1.12 IEEE 1451 Smart transducer block diagram that includes Smart Transducer Interface Module (STIM) with Transmission Electronic Data Sheet (TEDS) and Network Capable Application Processor (NCAP)

and the NCAP. It allows a variety of devices to have same hardware interface to the microprocessor. Figure 1.12 shows a system block diagram with the IEEE 1451.1 and 1451.2 interfaces.

The IEEE 1451.3 specifies the interface between the NCAP and smart transducers and TEDS for multi-transducers structure connected to the bus. The standard allows variety of sensors and actuators to be connected to the same NCAP through the bus structure, including both low and fast sampling rate sensors and actuators. IEEE 1451.4 deals with analog transducers and how they can be interfaced with microprocessors. The standard specifies TEDS connection for analog devices. The network can access TEDS data through digital communication first, and then send analog data to the analog actuator, for example. IEEE 1451.5 specifies a transducer to NCAP interface and TEDS for wireless communication scenarios. Common wireless communication protocols are included as transducer interfaces. The NCAP can then be implemented on some of the wireless devices and not physically attached to the sensor or actuator. IEEE 1451.7 defines interfaces for transducer-to-RFID (Radio Frequency Identification) systems.

Questions and Exercises

1. Describe Aloha protocol. What is the difference between Pure Aloha and Slotted Aloha protocols?
2. What are important design factors when wireless sensor networks are considered?
3. Describe one military application that uses wireless sensor networks. Can you think of a novel military application that uses the power of distributed sensing?

4. Research and articulate your own idea about a novel WSN application. Ask
   yourself who would buy such product/application and why? Research if there
   exist already a similar application using WSNs.
5. What are Medium Access Control (MAC), TDMA, FDMA?
6. What are specifics of an S-MAC protocol?
7. Describe basics of ZibBee protocol. What is the difference between ZigBee and
   IEEE 802.15.4 protocol?
8. What is the IEEE 1451 standard used for and why it is developed originally?
9. Describe chemical agents monitoring application and how cantilever-based
   sensors can be interfaced with wireless sensor networks?
10. What is a difference between wireless sensor node and the base station?

References

1. N. Abramson, "The Aloha System – Another alternative for computer communications,"
   Proc. AFIPS Joint Computer Conferences, 1970.
2. F. Bennett, D. Clarke, J.B. Evans, A. Hopper, A. Jones, and D. Leask, "Piconet: embedded
   mobile networking," *IEEE Personal Communications Magazine*, vol. 4, no. 5, pp. 8–15, Oct.
   1997.
3. E.H. Callaway, *Wireless Sensor Networks: Architectures and Protocols*, CRC Press LLC,
   Boca Raton, FL, 2004.
4. B. Chen, K. Jamieson, H. Balakrishnan, and R. Morris, "Span: an energy-efficient
   coordination algorithm for topology maintenance in ad hoc wireless networks," *Proc. 7th
   ACM International Conference on Mobile Computing and Networking*, Rome, Italy, July
   2001.
5. TinyOS Community Forum at http://www.tinyos.net/.
6. L. Doherty and K.S.J. Pister, "Scattered data selection for dense sensor networks," *Proc. the
   Third International Symposium on Information Processing in Sensor Networks*, April 26–27,
   2004, Berkeley, California, USA.
7. A. El-Hoiydi, "Aloha with preamble sampling for sporadic traffic in ad hoc wireless sensor
   networks," *Proc. IEEE International Conference on Communications (ICC 2002)*, New York,
   USA, pp. 3418–3423, April 2002.
8. D. Estrin, R. Govindan, J. Heidemann, and S. Kumar, "Next century challenges: Scalable
   coordination in sensor networks," *Proc. ACM/IEEE International Conference on Mobile
   Computing and Networking*, pp. 263–270 Seattle, Washington, August, 1999.
9. F. Figueroa and J. Schmalzel, "Rocket Testing and Integrated System Health Management",
   in *Condition Monitoring and Control for Intelligent Manufacturing* (Eds. L. Wang and R.
   Gao), pp. 373–392, Springer Series in Advanced Manufacturing, Springer Verlag, UK, 2006.
10. D. Jethwa, R.R. Selmic and F. Figueroa, "Real-time implementation of intelligent actuator
    control with a transducer health monitoring capability," *Proc. 16th Mediterranean
    Conference on Control and Automation*, Corsica, France, June 25–27, 2008.
11. H.-F. Ji, K.M. Hansen, Z. Hu, T. Thundat, "An Approach for Detection pH using various
    microcantilevers," *Sensor and Actuators*, 2001, 3641, 1–6.
12. X. Jiang, J. Polastre, and D. Culler, "Perpetual environmentally powered sensor networks,"
    IEEE Information Processing in Sensor Networks, 2005, pp. 463–468.
13. R.N. Johnson, "Building plug-and-play networked smart transducers," *National Institute of
    Standards and Technology IEEE 1451 Website*.

14. A. Kansal, J. Hsu, S. Zahedi, and M.B. Srivastava, "Power management in energy harvesting sensor networks," *ACM Transactions on Embedded Computing Systems*, 2006.
15. S. Kim, S. Pakzad, D. Culler, J. Demmel, G. Fenves, S. Glaser, and M. Turon, "Health Monitoring of Civil Infrastructures Using Wireless Sensor Networks," *Proc. the 6th International Conference on Information Processing in Sensor Networks (IPSN '07)*, Cambridge, MA, April 2007, ACM Press, pp. 254–263.
16. P. Kuckertz, J. Ansari, J. Riihijarvi, P. Mahonen, "Sniper fire localization using wireless sensor networks and genetic algorithm based data fusion," *IEEE Military Communications Conference*, Oct. 2007.
17. A. Ledeczi, A. Nadas, P. Volgyesi, G. Balogh, B. Kusy, J. Sallai, G. Pap, S. Dora, K. Molnar, M. Maroti, and G. Simon, "Countersniper system for urban warfare," *ACM Transactions on Sensor Networks*, vol. 1, no. 2, Nov. 2005, pp. 153–177.
18. K. Lee, "Brief description of the family of IEEE 1451 standards," *National Institute of Standards and Technology*, IEEE 1451 Website (cited July 2010): http://ieee1451.nist.gov/1451Family.htm.
19. F.L. Lewis, "Wireless Sensor Networks," book chapter in *Smart Environments: Technologies, Protocols, and Applications*, ed. D.J. Cook and S.K. Das, John Wiley, New York, 2004.
20. A. Mainwaring, J. Polastre, R. Szewczyk, and D. Culler, "Wireless sensor networks for habitat monitoring," *Intel Research*, June 2002.
21. M. Maroti, B. Kusy, G. Simon, and A. Ledeczi, "The flooding time synchronization protocol," *Proc. ACM Second International Conference on Embedded Networked Sensor Systems*, pp. 39–49, Baltimore, MD, November 3, 2004.
22. C.S.R. Murthy and B.S. Manoj, *Ad Hoc Wireless Networks: Architectures and Protocols*, Prentice Hall, Upper Saddle River, NJ, 2004.
23. M.A. Perillo and W.B. Heinzelman, "Wireless sensor network protocols," in *Algorithms and Protocols for Wireless and Mobile Networks*, Eds. A. Boukerche et al., CRC Hall Publishers, 2004.
24. J.G. Proakis, *Digital Communication*, McGraw-Hill, New York, NY, 2001.
25. M. Rahimi, H. Shah, G.S. Sukhatme, J. Heidemann, and D. Estrin, "Studying the feasibility of energy harvesting in a mobile sensor network," *Proc. IEEE International Conference on Robotics and Automation (ICRA)*, 2003.
26. T.S. Rappaport, *Wireless Communications*, Principles and Practice, Prentice Hall, Upper Saddle River, NJ, 2nd edition, 2002.
27. S. Senturia, *Microsystems Design*, Kluwer Academic Publishers, Norwell, MA, 2001.
28. K. Sohrabi and G.J. Pottie, "Performance of a novel self-organization protocol for wireless ad hoc sensor networks," *Proc. IEEE 50th Vehicular Technology Conference*, 1999.
29. C.M. Vigorito, D. Ganesan, A.G. Barto, "Adaptive control of duty cycling in energy-harvesting wireless sensor networks," *4th Annual IEEE Communications Society Conference on Sensor, Mesh and Ad Hoc Communications and Networks*, SECON '07, pp. 21–30, June 2007.
30. B. Warneke, M. Last, B. Liebowitz, K.S.J. Pister, "Smart Dust: Communicating with a Cubic-Millimeter Computer," *IEEE Computer*, January 2001.
31. G. Werner-Allen, J. Johnson, M. Ruiz, J. Lees, M. Welsh, "Monitoring volcanic eruptions with a wireless sensor network," *Proc. the Second European Workshop on Wireless Sensor Networks* (EWSN'05), 2005.
32. B. West, P. Flikkema, T. Sisk, and G. Koch, "Wireless sensor networks for dense spatio-temporal monitoring of the environment: a case for integrated circuit, system, and network design," *2001 IEEE CAS Workshop on Wireless Communications and Networking*, Notre Dame, Indiana, August 2001.
33. W. Ye, J. Heidemann, and D. Estrin, "An energy-efficient MAC protocol for wireless sensor networks," *Proc. IEEE INFOCOM*, New York, June 2002.
34. W. Ye, J. Heidemann, and D. Estrin, "Medium access control with coordinated adaptive sleeping for wireless sensor networks," *IEEE/ACM Transactions on Networking*, vol. 12, no. 3, June 2004.

35. F. Zhao and L. Guibas, *Wireless Sensor Networks*, Elsevier, 2004.
36. T. Zheng, S. Radhakrishnan, and V. Sarangan, PMAC: An adaptive energy-efficient MAC protocol for Wireless Sensor Networks," *Proc. IEEE International Parallel and Distributed Processing Symposium*, 2005.
37. W. Zhou, A. Khaliq, Y. Tang, H.-F. Ji, and R.R. Selmic, "Simulation and design of piezoelectric microcantilever chemical sensors," *Sensors and Actuators A*, vol. 125, no. 1, pp. 69–75, October 2005.
38. L.Q. Zhuang, K.M. Goh and J.B. Zhang, "The Wireless Sensor Networks for Factory Automation: Issues and Challenges," *Proc. IEEE Conference on Emerging Technologies and Factory Automation*, September 2007.
39. ZigBee Specification, publication by ZigBee Alliance, http://www.zigbee.org.
40. www.atmel.com.

Chapter 2
Topology, Routing, and Modeling Tools

In this chapter we discuss basic topology and routing concepts in WSNs, as well as mathematical modeling tools such as Voronoi diagrams and Delaunay triangulations that are used in setting up a framework for coverage, localization, and routing in WSNs.

2.1 Topology and Routing Protocols in WSNs

2.1.1 Topology in WSNs

The topology of a WSN refers to how the nodes are arranged within the network. Although wireless sensor networks consist of sensors that are miniaturized, pervasive, and coordinated, the general principles of topology of WSNs are the same as for any other network. The most common topologies are star, mesh and star-mesh hybrids topologies (see Fig. 2.1). Brief details of each of these topologies follow.

Star Topology In this topology the nodes are organized in the form of a star with the base station as the hub of the star. Sensor nodes broadcast data through the base station, and cannot directly exchange messages between each other. This topology offers low power usage as compared to other wireless sensor topologies. However, the base station cannot communicate with a node that is out of range. The fact that this topology depends on a single node to manage the network exposes it to a single-point-of-failure weakness, which negatively impacts the overall reliability of the network, i.e., the network is not very robust to failures of individual nodes.

© Springer International Publishing AG 2016
R.R. Selmic et al., *Wireless Sensor Networks*,
DOI 10.1007/978-3-319-46769-6_2

Fig. 2.1 Star topology: all
nodes directly connect with
the base station and thus
communicate with the rest of
network

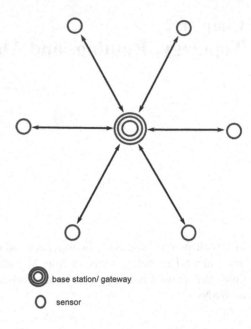

base station/ gateway

sensor

Mesh Topology A WSN with a mesh topology (see Fig. 2.2) has sensor nodes that communicate data through each other. This means that, if a sensor node wishes to send data to an out-of-range node, it can use another node as an intermediate communication resource. One advantage of this topology is that if a sensor node fails, communication is possible with other nodes that are within the communication range. A major disadvantage is that this topology uses more power due to redundant data transmission.

Star-Mesh Hybrid Topology A WSN with a star-mesh topology has attributes of both the star and mesh topologies. On one hand, this topology takes advantage of the low power consumption present in the star topology, while on the other, it takes advantage of the data redundancy present in the mesh topology to ensure data reaches its destination. In the implementation of this topology, nodes at the edge of the network are usually low-energy nodes, while nodes at the heart of the mesh have higher power (and could in some cases be plugged into the electrical mains), since they typically forward messages between large numbers of nodes and serve as a gateway nodes (Fig. 2.3).

2.1.2 Routing Protocols in WSNs

A routing protocol outlines how data is broadcasted through the network. Most routing protocols can be classified as data centric, hierarchical, location based, or QoS aware [1]. Brief details of each of these types of protocols follow.

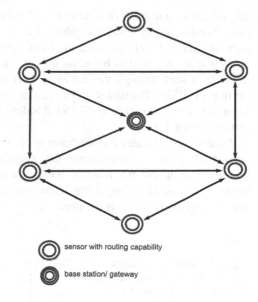

Fig. 2.2 Mesh topology: nodes do not have to directly connect to the base station, as they can communicate with it via other nodes

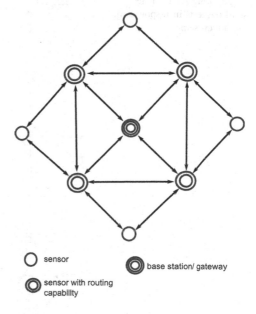

Fig. 2.3 Star-mesh topology: sensors with routing capabilities are connected in a mesh, such that regular sensors can communicate with the base station

Data Centric Protocols In large-scale WSN applications, the large number of randomly deployed nodes makes it infeasible to query sensors using their individual identifiers. One approach to addressing this problem is by sending queries to particular regions (set or cluster of sensor nodes) [1], such that data from sensors in that region is sent in response to the query. The challenge with this approach though is that data from a number of sensors in a given region contains a lot of redundancies, since sensors in any given neighborhood are likely to be sensing the same event (sensor data is highly correlated). Data centric protocols exploit attribute-based naming to aggregate data based on the data properties to eliminate redundancies as the data is sent through the network. This approach achieves significant energy savings in WSNs. Examples of data centric protocols include, Sensor Protocols for Information via Negotiation (SPIN), flooding and gossiping, directed diffusion and rumor routing [1].

Figure 2.4 illustrates the mechanism of operation of the SPIN protocol. First, a sensor advertises its data using the advertisement (ADV) message. Interested neighbors then use the request (REQ) message to request data. Following the request, data is then sent to the interested neighbors. The querying process continues recursively through the network.

Fig. 2.4 Mechanism of SPIN protocol: a node having data advertises this data, and then sends it to interested neighboring nodes if they send requests in response to the advertisement

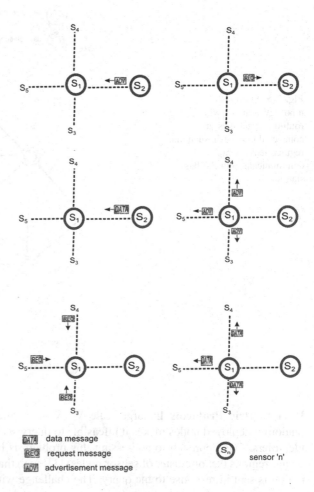

In the *flooding protocol*, each sensor broadcasts a received packet to all its neighbors until the packet reaches the destination. To avoid infinite looping, the packet propagation process may be interrupted if the packet exceeds a certain predetermined number of hops. Gossiping seeks to improve on the flooding protocol's extensive usage of resources, e.g., energy, due to the large number of messages moving around, by only advertising a received packet to a randomly selected neighbor.

In *directed diffusion*, the base station broadcasts data requests that are recursively sent through the network [2]. On receiving the requests, sensors nodes recursively set up gradients to the requesting nodes, until the gradients propagate back to the base station. A gradient is essentially a link to the requesting node, and defines the data rate, duration and expiration time associated with the request among other variables [1, 2]. In the final step, the best path is selected before data transfer begins.

Rumor routing improves upon directed diffusion by only routing queries to nodes that have sensed a particular event, as opposed to recursive propagation of requests to a wide range of nodes. To achieve this, the rumor routing uses agents, which are packets that convey information about events occurring at different locations of the network.

Hierarchical Protocols In this routing paradigm, illustrated in Fig. 2.5, the WSN is partitioned into clusters whose heads mainly perform tasks of processing (e.g., aggregation) and information forwarding, while the other nodes perform the sensing tasks within clusters. Hierarchical protocols have the advantage of being scalable due to the multi-tiered design while attaining high-energy efficiencies. Examples of hierarchical protocols include low-energy adaptive clustering hierarchy (LEACH), power-efficient gathering in sensor information systems (PEGASIS) and threshold sensitive energy efficient sensor network protocol (TEEN) [1], among others. A brief description of some of the main WSN hierarchical protocols follows.

In the *LEACH protocol*, the role of a cluster head rotates between sensor nodes to prevent a scenario in which the energy reserves of a few nodes may be drained at a much higher rate than the rest of the nodes. With a certain probability (which depends on the amount of energy left at the node), nodes elect themselves to be cluster heads. These cluster heads broadcast their status throughout the network, with the rest of the nodes assigning themselves to certain clusters depending on which cluster head's location is on a path requiring the least communication energy. The cluster head creates a schedule for the sensors in its cluster (e.g., when to turn radio on or off), aggregates data received from nodes within the cluster and also transmits the aggregated data to the base station.

PEGASIS protocol improves on the performance of LEACH by having nodes communicate only with their immediate neighbors, with a single designated node transmitting the data to the base station in each round. To minimize the average amount of energy spent by each node per round, the task of transmitting data to the base station is taken up by different nodes in turns.

Fig. 2.5 A hierarchical
clustering example: sensor
nodes are clustered around
first-level cluster heads, which
in turn communicate to the
base station via second-level
cluster heads

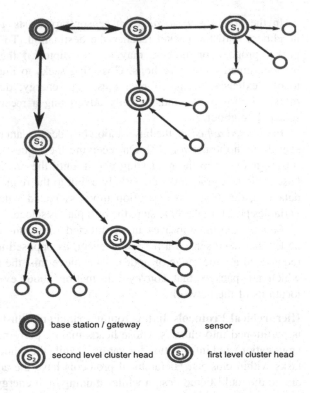

TEEN protocol is an energy-efficient WSN routing protocol that is designed for
time-critical WSN applications in which changes in the sensed variable require
immediate reaction. Nodes continuously perform the sensing functions, with the
messages broadcast from the base station including threshold values that are used as
basis for triggering sensors to forward data that has been sensed. TEEN is not
suitable for applications that require data to be continuously relayed, since sensors
will never transmit if thresholds are not exceeded.

Location Based Protocols These protocols use information about sensor location
to route data in an energy-efficient way [1]. The distance between two sensor
locations is calculated and its energy requirement estimated. Location based protocols include minimum energy communication network (MECN), geographic
adaptive fidelity (GAF) and geographic and energy aware routing (GEAR) [1].

MECN is applicable to both WSNs and mobile ad hoc networks, and uses GPS
to keep track of node locations. The protocol uses node-positioning information to
identify paths in the network that minimize the energy required for data transfer.
The protocol is built on the concept of a relay region for each node, which defines a
set of neighboring nodes via which transmission is cheaper (in energy terms) than
direct transmission between a given node and the destination.

GAF was designed for mobile ad hoc networks, but can also be used with WSNs.
GAF uses location information to turn off or turn on certain nodes for energy

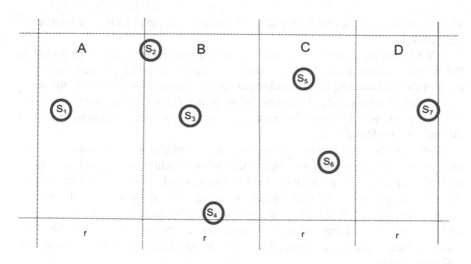

Fig. 2.6 Virtual grid in a GAF protocol: sensors in the same grid can alternate between active and passive states for load balancing and energy conservation

conservation purposes, in such a way not to compromise the routing tasks. Figure 2.6 shows a simple illustration of the function of the GAF protocol. Each node uses its GPS location to associate itself with a point on the virtual grid. Nodes belonging to the same grid are regarded as redundant and alternate between active and dormant states in a bid to save energy. In Fig. 2.6 sensors 2, 3, and 4 are all in the same grid. Similarly sensors 5, and 6 are also located in the same grid. Sensor 1 can reach any of sensors 2, 3, or 4 in region B, and any of them can reach either of sensor 5 and 6 in region C. Therefore, to save energy, only one sensor in region B and C will be active at any given moment while the others are dormant. The sensors will alternate states to balance the load between all sensors in the associated region.

The *GEAR* protocol uses sensor location information to improve on the energy usage of the generalized directed diffusion process by sending queries to only a few selected regions of the network (instead of the whole network). This kind of selective querying is especially useful in sensor network applications where the aim is to collect data on location basis (e.g., the temperature at a certain point at a certain time).

QoS Aware Protocols The previously described categories of WSN routing protocols could also be considered to be QoS-aware, since they seek to optimize variables such as energy consumption, a fundamental factor in determining the QoS obtained from a WSN. As such, the definition of a QoS-aware protocol is not well streamlined in WSN literature, and we briefly discuss the routing protocols which touch on QoS elements such as end-to-end delays and prioritization of packets in the network.

SPEED (*not an acronym*) is a routing protocol that ensures that every packet in the network attains a certain speed during a transfer [1]. The protocol utilizes a form of data acknowledgement mechanism that enables a node to estimate the delay to a given neighbor, which in turn helps to determine which route meets the required transmission speeds. The protocol also offers a form of congestion control during a network overload.

SAR (*sequential assignment routing*) uses the priority of a given packet, network energy and QoS considerations along a link while making routing decisions. Each node is associated with a tree (rooted at the node), which defines a set of paths from the node in question. The paths generally avoid low-energy nodes and are periodically recomputed. The multi-path approach used by SAR makes it fault tolerant, with the best of the tabulated paths being used for routing. For large WSNs, the protocol faces significant overhead in maintaining state and all the tables at the different sensors.

Energy-aware QoS routing protocol is another routing protocol that seeks to minimize energy consumption and end-to-end delays. The protocol classifies traffic as best effort or real-time (real-time data being mainly data from imaging sensors), with real-time data seeing higher priority during times of heterogeneous traffic. The overall performance of the protocol depends on the setting of the bandwidth ratio, a parameter that determines the bandwidth share between different traffic types during congestion.

Other WSN protocols that have been classified as QoS-aware include maximum lifetime energy routing, maximum lifetime data gathering and minimum cost forwarding among others [1].

2.2 Modeling Tools

2.2.1 Voronoi Diagrams

A Voronoi diagram, such as the one shown in Fig. 2.7, is one of the most useful structures in computational geometry. It is named after Georgy Voronoi, a Russian mathematician who generalized and formally defined the *n*-dimension case in [21]. The origin of Voronoi diagram dates back to Descartes [9] in the 17th century. The first computational algorithm for constructing Voronoi diagram is presented by Shamos and Hoey [19], who investigate in problems regarding to the proximity of a finite set of distinct points in Euclidean space, such as finding a minimum spanning tree, identifying the smallest circle enclosing the set of points, locating the nearest and farthest neighbors. Ever since, a lot of researchers have devoted to this field. A dual diagram that [23] can be constructed from Voronoi diagram is called Delaunay tessellation or Delaunay triangulation, which is named after Boris Delaunay [8]. A complete overview on the Voronoi diagrams can be found in [18].

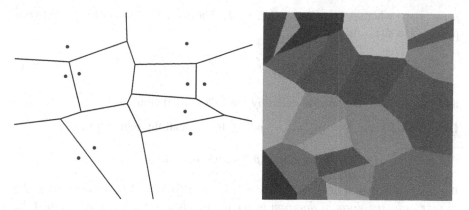

Fig. 2.7 Two examples of Voronoi diagrams created with randomly deployed points

Fig. 2.8 The bisector of two
sites divides the plane into
two half-planes

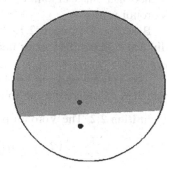

Voronoi diagram is a form of decomposition of metric space with respect to the Euclidean distances (other metrics can also be used) to specific set of points in the space. We consider only the situation when there are finite points in the space. Voronoi diagram with infinite number of points are called infinite Voronoi diagrams and are described in [18].

Let S be a set of distinct points $p_1, p_2, \ldots, p_n$ in the Euclidean space, where $n < \infty$ and $I_n = \{1, 2, \ldots, n\}$. To distinguish points p_i' that $p_i' \in S$ and point p_i that $p_i \in S$, we define point p_i as Voronoi site p_i [3], or site p_i for short. In case of 2-dimensional Euclidean space, each of the generated Voronoi polygons will contain only one site $p_i \in S, i \in I_n$. Meanwhile, any site $p_i' \notin S$ that is inside a give polygon $V(p_i)$ is closer to the corresponding site $p_i \in S$ than to other sites in S. Voronoi diagram divides the plane into regions such that each point of a single region has the same closest site, Fig. 2.8. The formal mathematical definition of a planar ordinary Voronoi diagram is stated in [18]. Suppose the Cartesian coordinates of site p_i are labeled as a pair

$p_i = (x_i, y_i)$ where $p_i \neq p_j$ for $i \neq j$, $i, j \in I_n$. Then, the Euclidean distance between point p and site p_i is

$$d(p, p_i) = \sqrt{(x - x_i)^2 + (y - y_i)^2}. \tag{2.1}$$

and the Voronoi diagrams is given by the following definition:

Definition 2.1 Let $S = \{p_1, p_2, \ldots, p_n\} \subset \Re^2$, we call the region given by

$$V(p_i) = \{p \| p - p_i \| < \| p - p_j \| \quad \text{for } i \neq j\}. \tag{2.2}$$

the Voronoi polygon of p_i. The set $V = \{V(p_1), V(p_2), \ldots, V(p_n)\}$ is called as the *planar ordinary Voronoi diagram* generated by sites S. Each site p_i is called the *generator site* of the corresponding Voronoi polygon $V(p_i)$, Fig. 2.9. The boundaries of Voronoi polygon, which is defines as *Voronoi edges*, consist of line segments, half lines or infinite lines. The end points of Voronoi edges are called Voronoi vertices.

Based on Definition 2.1, the bisector of any pairwise site i and site j $(i \neq j)$ divides the Euclidian space into two half spaces, $H(p_i p_j)$ and $H(p_j p_i)$ as shown in Fig. 2.8.

The Voronoi polygon $V(p_i)$ can be formed as an intersection area of half spaces, which is considered in the following definition.

Definition 2.2 The Voronoi polygon $V(p_i)$ can be formed as

$$V(p_i) = H(p_i p_1) \cap H(p_i p_2) \cap \cdots \cap H(p_i p_n). \tag{2.3}$$

Suppose there are four sites in Euclidian space. Figure 2.10 shows the process to form the Voronoi polygon $V(p_1)$ based on Definition 2. 2.

Voronoi polygons of other sites can also be formed in the same fashion and the computation complexity is $O(n^2 \log n)$. Shamos [20] have shown that a lower bound for the computation of Voronoi diagrams is $O(n \log n)$, where x-coordinates of all sites have to be strictly increased. Some algorithms such as divide and conquer

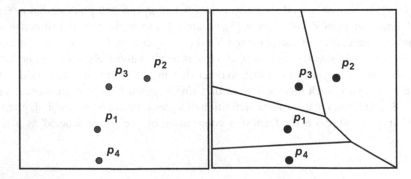

Fig. 2.9 The Voronoi generator site and the corresponding Voronoi polygon

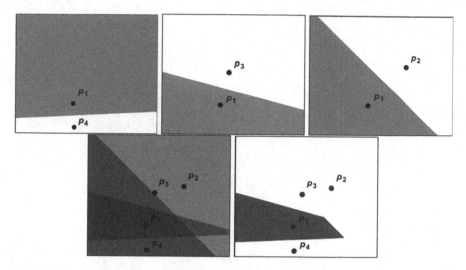

Fig. 2.10 The formation of Voronoi polygon as described in Definition 2.1

based algorithm and sweep line based algorithm can run in $O(n \log n)$ with specific data structure to satisfy the same condition. For example, Fortune's algorithm [10] uses binary search tree to store the structure of sweep line.

Voronoi diagrams and Delaunay triangulations have been widely used in many fields [3, 4, 11, 18] from biology to chemistry, from marketing to astronomy, and many others. In the area of wireless sensor networks, Voronoi diagrams have been extensively used for coverage related problems [14, 15, 22], motion planning problems [16, 24], routing problems [12, 17], localization problems [5], target tracking [7], and others.

2.2.2 Delaunay Triangulations

A triangulation T on a point set P in two-dimensional space (can also be equivalently analyzed in higher-dimensional spaces) is a set of triangles such that [6]:

a. A point p is a vertex of a triangulation triangle if and only if $p \in P$;
b. The intersection of two triangles is either an empty set or an edge of a triangle;
c. The set T is maximal: there does not exist any triangle that can be added to T without violating the previous rules.

A triangulation T is a Delaunay triangulation if and only if the circle circumscribing each triangle does not contain any point of the set P. Figure 2.11 shows Delaunay triangulation for a set of points in two-dimensional space.

Note that Delaunay triangulation is not unique if there are four or more points lying on the same circle. For example, consider a square and its four vertices. There

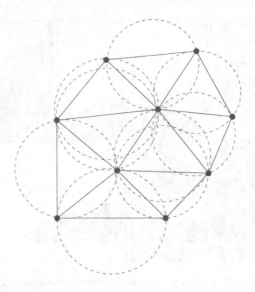

Fig. 2.11 Delaunay triangulation for a set of points in two-dimensional space

are two possible ways to triangulate those four points (draw those triangulations as
an exercise). In case of all points lying on the same line, triangulation is not
possible.

Delaunay triangulation and Voronoi diagrams are dual. Two vertices are connected in Delaunay triangulation if and only if they share a common boundary in
Voronoi diagram. The Voronoi vertex is the circumcenter of some Delaunay triangle, Fig. 2.12.

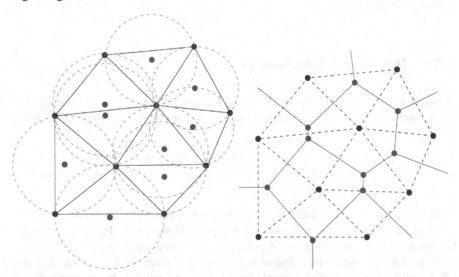

Fig. 2.12 Delaunay triangulation is dual to Voronoi diagram: circumscribing *circles* and their
centers (*left*) and Voronoi diagram created from the Delaunay vertices (*right*)

Delaunay triangulation or Voronoi diagram cannot be generated in a fully distributed fashion. For a large circumcircle of three nodes, one needs to check if the circle is empty of other nodes [13], thus requiring centralized information such as location of all other nodes in the network. Construction of Delaunay triangulation of full network would require extensive communication among all nodes, and therefore, is not very practical. Its construction time complexity is $O(n \log n)$ [13].

Questions and Exercises

1. Briefly describe the concept of data aggregation as used in wireless sensor networks. What are its advantages? You may use a drawing and (or) an example of a WSN application of your choice for illustration purposes.
2. One of the categories of routing protocols for WSNs is a group of QoS-aware protocols. How do you understand the term "QoS" as used in WSNs? How do the QoS demands of WSNs differ from those of regular computer networking applications?
3. Describe in detail the mechanism of operation of one QoS-aware routing protocol in WSNs. Can a QoS-aware routing protocol be hierarchical at the same time? Please explain.
4. What is the SPIN protocol's major weakness? For what kinds of applications would SPIN be unsuitable? With the aid of diagrams if possible, briefly describe the weaknesses and strengths of the flooding and gossiping protocols for WSNs.
5. Given a set of points $A = \{(-2, 2), (2, 2), (-2, 2), (-3, -3)\}$, draw the Voronoi diagrams and Delaunay triangulations. Repeat the same task for the set $B = \{(-2, 2), (2, 2), (-2, 2), (-2, -2)\}$.
6. Explain why Voronoi diagrams and Delaunay triangulations are dual? How is such duality related to their geometrical interpretation?
7. Given N homes in a region, where within the region to place a nuclear power plant such that it is as far away from any home as possible?

References

1. K. Akkaya and M. Younis, "A survey on routing protocols for wireless sensor networks," *Ad Hoc Networks*, vol. 3, pp. 325–349, 2005.
2. J.N. Al-Karaki and A.E. Kamal, "Routing techniques in wireless sensor networks: a Survey," *IEEE Wireless Communications*, vol. 11, pp. 6–28, 2004.
3. F. Aurenhammer and R. Klein, "Voronoi diagrams," in J. Sack and G. Urrutia, editors, *Handbook of Computational Geometry*, Elsevier Science Publishing, 2000.
4. F. Aurenhammer, "Voronoi diagrams: A survey of a fundamental geometric data structure," *ACM Computing Surveys*, vol. 23, pp. 345–405, 1991.

5. A. Boukerche, H. Oliveira, E. Nakamura, and A. Loureiro, "Dv-loc: a scalable localization protocol using Voronoi diagrams for wireless sensor networks," *IEEE Wireless Communications*, vol. 16, no. 2, pp. 50–55, April 2009.
6. P. Cignoni, C. Montani, and R. Scopigno, "DeWall: A fast divide & conquer Delaunay triangulation algorithm in E^d," *Computer-Aided Design*, vol. 30, no. 5, April 1998, pp. 333–341.
7. W.P. Chen, J. Hou, and L. Sha, "Dynamic clustering for acoustic target tracking in wireless sensor networks," *IEEE Transactions on Mobile Computing*, vol. 3, no. 3, pp. 258–271, July-August 2004.
8. B. Delaunay, "Sur la sphère vide," *Otdelenie Matematicheskikh i Estestvennykh Nauk*, vol. 7, pp. 793–800, 1934.
9. R. Descartes, "Principia Philosophiae," *Ludovicus Elzevirius*, Amsterdam, 1644.
10. S. Fortune, "A sweepline algorithm for Voronoi diagrams," *Proc. the 2nd Annual Symposium on Computational Geometry*, pp. 313–322, 1986.
11. M. Held, "On the computational geometry of pocket machining," *Lecture Notes in Computer Science,* Springer-Verlag, 1991.
12. J.S. Li, H.C. Kao, and J.D. Ke, "Voronoi-based relay placement scheme for wireless sensor networks," *IET Communications*, vol. 3, no. 4, pp. 530–538, April 2009.
13. X.-Y. Li, P.-J. Wan, and O. Frieder, "Coverage in wireless ad hoc sensor networks," *IEEE Transaction on Computers*, vol. 52, no. 6, June 2003.
14. S. Megerian, F. Koushanfar, M. Potkonjak, and M. Srivastava, "Worst and best-case coverage in sensor networks," *IEEE Transactions on Mobile Computing*, vol. 4, no. 1, pp. 84–92, January–February 2005.
15. S. Meguerdichian, F. Koushanfar, M. Potkonjak, and M. Srivastava, "Coverage problems in wireless ad-hoc sensor networks," *Proc. 20th Annual Joint Conference of the IEEE Computer and Communications Societies (INFOCOM 2001)*, vol. 3, pp. 1380–1387, 2001.
16. T. Melodia, D. Pompili, and I. Akyildiz, "A communication architecture for mobile wireless sensor and actor networks," *3rd Annual IEEE Communications Society on Sensor and Ad Hoc Communications and Networks*, vol. 1, pp. 109–118, September 2006.
17. T. Melodia, D. Pompili, and I. Akyldiz, "Handling mobility in wireless sensor and actor networks," *IEEE Transactions on Mobile Computing*, vol. 9, no. 2, pp. 160–173, February 2010.
18. A. Okabe, B. Boots, K. Sugihara, and S. N. Chiu, *Spatial Tessellations-Concepts and Applications of Voronoi Diagrams*, John Wiley, Second Edition, 2000.
19. M.I. Shamos and D. Hoey, "Closest-point problems," *Proc. of the 16th IEEE Symposium on Foundations of Computer Science*, pp. 151–162, 1975.
20. M.I. Shamos, Computational Geometry, *Ph.D. Dissertation*, Yale University, 1978.
21. G.M. Voronoi, "Nouvelles applications des paramètres continus à la théorie des formes quadratiques," *Journal für die Reine und Angewandte Mathematik*, vol. 133, pp. 97–178, 1908.
22. G. Wang, G. Cao, and T. La Porta, "Movement-assisted sensor deployment," *Proc. 23rd Annual Joint Conference of the IEEE Computer and Communications Societies*, vol. 4, pp. 2469–2479, March 2004.
23. E.W. Weisstein, "Dual graph," From MathWorld—A Wolfram Web Resource, http://mathworld.wolfram.com/DualGraph.html.
24. A. Yazici, G. Kirlik, O. Parlaktuna, and A. Sipahioglu, "A dynamic path planning approach for multi-robot sensor-based coverage considering energy constraints," *International Conference on Intelligent Robots and Systems*, October 2009.

Chapter 3
WSN Architecture

A wireless sensor network (WSN) is a system that consists of multiple sensing elements distributed spatially with a specific objective to measure different physical quantities and communicate those measurements between themselves and the central gateway. These sensing elements (sensors) are used to observe and instrument several physical and environmental conditions such as motion, pressure, temperature, sound, etc. [6, 38].

Sensor nodes are deployed in a domain or area, which is called a sensing field. Each of these sensor nodes typically has the capability to collect data from the sensing field, and may in some cases analyze or route the data to its destination node. A sensor node typically consists of the following components (see Fig. 3.1 for arrangement of the elements):

- sensing and actuation unit (single element or array)
- processing unit (e.g., a microcontroller)
- communication unit (radio transceiver or other wireless communications device)
- power unit
- other application-dependent units.

Analog signals detected by the sensor are converted into digital format by an analog–digital converter (ADC), which then outputs the digital signal into the processing unit. The processing unit manages all the sensor's operational procedures, with the transceiver connecting the sensor to the rest of the network. Modules such as the location finding system and the mobilizer are application dependent. The former is needed where sensors may have to move during operation while the latter helps provide accurate location information [2]. The power module may include energy scavenging elements that derive energy from ambient heat, light, radio, and vibrations.

WSN architectures are governed by the applications and their requirements. In most cases, hardware and software requirements are small form-factor, efficient and smart energy usage, built-in redundancy, in situ reprogramming, and, most recently, a high security. The design of WSNs follows these requirements as will be shown in

© Springer International Publishing AG 2016
R.R. Selmic et al., *Wireless Sensor Networks*,
DOI 10.1007/978-3-319-46769-6_3

this chapter. In this chapter we also discuss fundamentals of hardware design and basic components of layered network architecture such as physical layer, data link layer and network layer. The hardware components are divided into sensor nodes and network controllers or base stations. We describe the most important hardware building blocks that are common for almost all sensor nodes as well as common sensors and underlying sensing principles.

3.1   Components of a Wireless Sensor Node

Sensor nodes consist of a variety of sensors (sometimes built on a separate module called a sensor module or sensor board), a microcontroller or microprocessor for on-board communication and signal processing, memory, radio transceiver with antenna for communication with neighboring nodes, power supply, and supporting circuitry and devices, Fig. 3.1. Most of the sensor nodes run their own operating system developed for small form-factor, low-power embedded devices, such as TinyOS [29], that provides inter-processor communication with the radio and other components in the system, controls power consumption, controls attached sensor devices, and provides support for network messaging and other protocol functions.

The choice of specific components depends on the application that the node is intended for. While the physical quantity to be measured determines required type of sensors, the selection of sensors in turn determines the interface type (such as analog-to-digital converter) between sensors/actuators and microcontroller/microprocessor. A signal processing algorithms, in-network data aggregation, and other application data processing determine the type of microcontroller or microprocessor that is needed. Applications where heavy processing and distributed, node-based data analysis is required necessitate microprocessors rather than smaller, yet more energy efficient microcontrollers. The choice of the central controller and data storage determines the type of memory device. The radio component is chosen based on the application requirements for data rate, range, bandwidth, communication

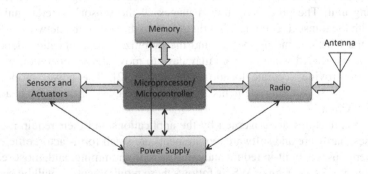

Fig. 3.1 Wireless sensor node architecture, the main building blocks

protocol selection and power budget. Power supply must have enough capacity to support node's components and small enough size to fit into the sensor node package that is again determined by the application requirements.

Here we review the basic hardware components, their functionality, and provide an overview of the most widely used components in today's WSN nodes. We cover basic sensors and actuators that are used in WSNs, microcontrollers and microprocessors, memory, radio with antenna, and the power supply.

3.1.1 Sensors and Actuators

Sensors are transducers that convert a physical quantity or a parameter into a signal with equivalent information [10, 42]. For instance, temperature sensors convert temperature into an electronic signal and further into digital number that users can read. The output signals in modern sensors are mostly electrical (information translated into the form of voltage, current, or charge) or optical (information translated into light intensity, wavelength, polarization, or phase). Depending on the measured quantity, sensors can be classified into mechanical sensors, thermal sensors, electrostatic and magnetic sensors, radiation sensors, chemical sensors and biological sensors [42]. A device with the opposite functionality is called an actuator where a signal is converted in some form of mechanical action or motion, for instance a servomotor that converts an electronic signal into motor shaft movement. Sensors can also be classified as *passive* or *active*, depending on their source of energy [2]. Passive sensors measure and sense without need for any additional energy sources. An example is a photodiode that passively measures light intensity without the need for additional stimulus or excitation. On the other hand active sensors require external stimulus or excitation to measure quantity. An example is a thermistor where a current passing through is required so that changes in resistance are reflected on a voltage across the thermistor. Obviously due to WSNs strict energy constraints, passive sensors are preferable, but there are also numerous WSNs applications that involve active sensors as well.

The most commonly used sensors in WSN applications and their related sensing principles are briefly described next.

Accelerometers Accelerometer sensors measure acceleration. Acceleration is a derivative of velocity, which is a derivative of displacement, or equivalently, acceleration is a second derivative of displacement, allowing displacement sensors to be used for measuring acceleration. Such methods can be used only for low frequency applications (bandwidth of 1 Hz or less, [10]) because measurement noise is amplified after differentiation calculation. For higher frequency sensing application a direct acceleration measurement is necessary. Common accelerometer

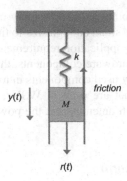

Fig. 3.2 Spring-mass-damper, a main principle of operation for accelerometers

sensors are based on spring-mass-damping dynamical system (Fig. 3.2) where a small mass, called seismic mass, is attached with springs to the body of the sensor. Force (acceleration) that is applied to the mass will cause a displacement or an oscillation of the seismic mass, and that in turn is related to the applied force (acceleration).

A dynamic equation that describes the spring-mass-damper is given by

$$M\frac{d^2y(t)}{dt^2} + F_{\text{friction}} + F_{\text{spring}} = r(t), \tag{3.1}$$

where $y(t)$ is a displacement and $r(t)$ is an applied force. Linear approximation of these forces yields

$$M\frac{d^2y(t)}{dt^2} + b\frac{dy(t)}{dt} + ky(t) = -Ma(t), \tag{3.2}$$

where b is a friction coefficient, k is a spring constant, and $a(t)$ is the input acceleration of the accelerometer body.

There are several methods for measuring displacement as a part of accelerometer sensor including capacitive, piezoresistive, piezoelectric, and thermal accelerometers. The capacitive accelerometers [15] measure displacement of capacitor plates that are results of spring-mass-damper model movement. They record either accumulated change, or distance between capacitor plates, or their overlapping area. Those changes will reflect on modified capacitance and then interface circuit converts capacitance to voltage changes. Three-dimensional micro-accelerometers have a seismic mass built in on the top of the silicon wafer and are reduced in size to around 1 μm. The mass is attached to the wafer using polysilicon springs,

Fig. 3.3 Capacitor-based accelerometer built on the top of the silicon wafer

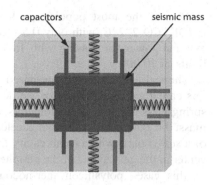

capacitors   seismic mass

Fig. 3.3, and deflection is measured using moving capacitor plates creating differential capacitor [2, 22]. The mass can have attached springs and capacitors in all three directions thus measuring acceleration, through capacitor's plate displacement, in all three dimensions. Often, multiple capacitors were placed forming differential capacitors that increases sensitivity and compensates for environment interferences.

Another type of accelerometer is a piezoresistive accelerometer that has strain gauges embedded into supporting springs. The strain gauges measure a change in resistance as a function of strain and that is related to the mass displacement, which in turn is related to the acceleration. These devices can be used to measure acceleration in a broad frequency range up to 13 kHz.

Piezoelectric accelerometers are based on piezoelectric effect that converts a force or strain to electric charge. Piezoelectric materials have the characteristic that when the strain is applied on them, there is an internal electric polarization in a form of aligned electric dipoles within the material resulting in a charge accumulated on the material surface [25]. This effect is reversible, meaning that the external charge will cause the strain in the material. Commonly used piezoelectric materials are quartz, gallium arsenide, zinc oxide, aluminum nitride, and lead zirconate-titanate (PZT). Usually, the piezoelectric component is placed between the sensor housing and the seismic mass; see Fig. 3.4 [20]. Acceleration causes the mass to apply the force on a piezoelectric material causing an electric charge accumulation, which is in turn fed into an interface circuitry for conversion into an useful sensor signal.

Fig. 3.4 Structure of a piezoelectric accelerometer

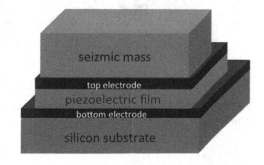

seizmic mass

top electrode

piezoelectric film

bottom electrode

silicon substrate

One of the most popular small form-factor piezoelectric accelerometers is ENDEVCO 2222C, with only 0.5 g of weight. It is a passive device requiring no external power for sensing with frequency range up to 10 kHz and acceleration limits of 10,000 g.

Thermal accelerometers are based on heat transfer between a seismic mass, in this case a gas, and the rest of the device. The principle is similar to the standard spring-mass-damping accelerometers, except that gas molecules serve as a seismic mass. An example of a thermal accelerometer is given in [10]. The sensor consists of a sealed plate with a small cavity filled with gas. A heat source is located in the center of the gas cavity. At the periphery of the cavity are four temperature sensors, in this case—polysilicon thermocouples, that measure temperature difference between these points (see Fig. 3.5).

When there is no acceleration, the heated gas is equally distributed inside the cavity, causing the zero temperature difference readings on the thermocouples. When there is an acceleration applied to the sensor, it causes heated molecules to shift opposite of the acceleration direction, thus generating the temperature gradient. The voltage reading on the thermocouples is proportional to the temperature difference $\Delta T = T_2 - T_1$ of the thermocouples, which in turn is proportional to the acceleration. With four temperature sensors distributed symmetrically across the gas cavity, the sensor can detect acceleration in x–y directions. An advantage of this thermal sensor is that it can withstand large mechanical shocks since it is gas operated. A drawback is that it is an active sensor that requires heater for its operation and is sensitive to the changes in ambient temperature.

Photodiode A photodiode is a photodetector (sensor that detects light) that is used to measure a light energy. The sensor generates current or voltage as a result of light energy that excites an electron in the photodiode, thus producing a free electron or a negative charge, and a hole or a positive charge. Such effect is called a photo effect, discovered by Albert Einstein who was awarded the Nobel Price for this discovery.

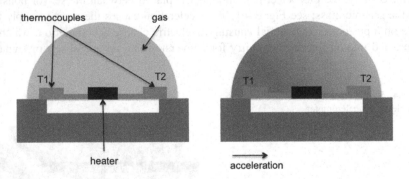

Fig. 3.5 Thermal accelerometer with gas as a seismic mass: no acceleration, $T_1 = T_2$ (*left*), and with acceleration where $T_1 > T_2$ (*right*)

Photocurrent is created, as free electrons move towards cathode and holes move towards anode of the sensor. Photodiodes are mostly made of semiconductor materials even though there are some attempts to fabricate them using polymers and other materials [35]. Semiconducting materials used in photodiodes fabrication include silicon (S_i), gallium arsenide (G_aA_s), indium antimonide (I_nS_b), indium arsenide (I_nA_s), and others. Photodiodes have p–n junction or p–i–n junction structure, Fig. 3.6. The p–n junction consists of p-type (large number of holes) and n-type (doped to produce a large number of electrons) semiconductors joined together forming a p–n junction, [25].

There is a concentration gradient that causes electrons to move into the p-layer and holes to move into the n-layer, generating a potential difference or a voltage across the junction. This potential field is opposed with diffusion forces for both electrons and holes. When there is not external bias voltage (external voltage between anode and cathode), the equilibrium state is reached when the electric field forces are equal to diffusion forces.

The photodiode converts the energy of light (photons) into free electrons and holes. When a photon hits the diode with sufficient energy, it converts some of that energy to higher energy state of the atoms, creating a loose or free electron and free hole (photoelectric effect). If such a process occurs in the depleted region with active electric field, the field will cause electrons to flow toward cathode, and holes toward anode creating the photocurrent.

Photodiodes are often created using a p–i–n junction structure where the i-region stands for an intrinsic semiconductor—a lightly doped semiconductor region that causes the diode to have small rectifying effect, but suitable for photodetectors. Typically, a light enters the photodiode through the p-region and photons are absorbed at the intrinsic region. A charge that is generated is moved across the junction by the internal bias that creates the electric field. This movement of charge across the junction represents a small photocurrent that can be detected at the diode electrodes. The size of the intrinsic region affects the quickness of the device response, i.e., larger the region, most of the charge carriers is created in the intrinsic region, increasing the efficiency of the device since there will be low level of recombination [16].

The electric symbol for the photodiode is shown in Fig. 3.7.

Fig. 3.6 Photodiode with p–n junction (*top*) and p–i–n junction structure (*bottom*)

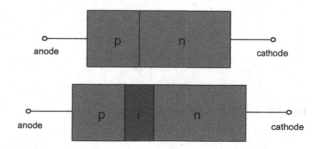

Fig. 3.7 Electric symbol for
a photodiode

Magnetometer A magnetometer is a sensor that measures magnetic field. There is
a plethora of these devices that are used in variety of applications including sensor
networks, navigation and control, control of aerial vehicles, drilling in oil and gas
industry, medical applications, cell phones, smart phones, e-readers, and many
others. Here we focus on magnetometers that are common in wireless sensor networks applications.

The Hall effect magnetometers are based on an effect discovered by Edwin Hall
in 1870—charges that are moving in the magnetic field experience the Lorentz
force, and therefore, move on a side of the conductor or semiconductor where they
flow, causing a potential difference that is proportional to the magnetic field.
Figure 3.8 illustrates the concept of the Hall effect sensor.

The force that electrons or holes experience while moving in the magnetic field
is given by

$$\vec{F} = q(\vec{E} + \vec{v} \times \vec{B}), \tag{3.3}$$

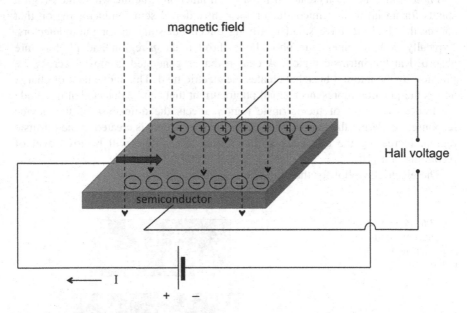

Fig. 3.8 Hall effect-based magnetometer

where $\vec{F}$ is the Lorentz force vector, q is the charge, $\vec{E}$ is the electric field vector, $\vec{v}$ is the velocity vector, and $\vec{B}$ is magnetic field vector. Hall effect magnetometers are vector-based magnetometers as they can measure not only magnitude, but are sensitive to the direction of the magnetic field as well. They are usually small sized, low-power, and with somewhat limited sensitivity.

A fluxgate is another type of magnetometer. It consists of an inductive coil that is wound around a magnetic core material [21]. When the permeability of the core material is changed, the flux in the core is changed, and as a result a voltage is induced in the coil. The induced voltage V is given by

$$V = nA \frac{dB}{dt}, \qquad (3.4)$$

where n is the number of turns in the coil, A is the cross-section area, and B is the core magnetic field. If a fluxgate has two oppositely magnetized cores inside the same coil, then the two magnetizations cancel each other, and the flux change is caused only by the external magnetic field. These types of sensors have two magnetic cores with two magnetizing coils that have opposite winding directions. Their magnetic fluxes cancel each other producing a zero voltage at the external (also called pick-up) coil. The fluxgate sensor measures the magnetic field in the direction of the coil. However, it takes three independent flux gate coils to measure total field by adding the three independent vectors. The sensitivity is limited due to high precision requirements in orientation of the three coils axes. Fluxgate advantage is the vector measurements—direction and magnitude of the magnetic field as well as a wide measurement range and the low noise level.

If direction is not required, then scalar magnetometers can be used. *Nuclear free-precession* is a scalar magnetometer that stimulates and polarizes the atomic nuclei of a substance causing the nuclei to spin, or precess, around a modified axis [18]. Nuclear precession magnetometers stimulate the atomic nuclei of a substance causing the nuclei to spin (the correct term is precess) temporarily around a new axis. When the excitation current (magnetic field) is turned off, the precessing protons generate a signal in the coil whose frequency is proportional to the strength of the magnetic field.

Magnetometers are used in a wide variety of applications within sensor networks. For example in undersea monitoring and surveillance, magnetometers are used to detect objects passing nearby the network, [31]. The system uses the Helium-3 nuclear precession total-field magnetometer to network with other nodes via acoustic communication channel, and to monitor for submarines and other undersea objects.

A small form factor magnetometer from Freescale (MAG31100 series, [43]) is commonly used in cell phones, smart phones, laptops, and sensor networks. Other common applications where magnetometers are used include compass, cell phones, GPS receivers, cars, airplanes, ships, and others.

Chemical and Gas Sensors Chemical agent monitoring systems are of great importance in detecting, identifying, locating, and quantifying specific chemical agents to give security officers an early warning to avoid contamination [4, 17].

A chemical sensor is a device that transforms quantitative or qualitative chemical information into a useful signal as a result of interaction between the gas and the sensor [12]. Exposure of the gas sensor to a gaseous chemical compound or a mixture of compounds results in a change of one or more of the physical properties of the sensor. This change is measured either directly or indirectly and it conveys information ranging from the concentration of a specific component to the total composition analysis of the ambient gas. A chemical sensor is classified based on its principle of operation or the physical property of the system that it measures and each class of the sensors has dissimilar selectivity and sensitivity. Properties of these sensors can usually be altered by varying the parameters such as sensing material or temperature. Chemical sensors consist of two fundamental functional units: namely, the receptor part and the transducer part. The receptor part of the sensor transforms the chemical information into energy that can be measured by the transducer and the transducer part changes this energy into a useful analytical signal. The receptor part of the sensor maybe classified as physical, chemical or biochemical based on the process that gives rise to the analytical signal. Sensors do not necessarily respond in a specific way to certain analytes present in different sample types. However, in well-defined conditions, the response is independent of other sample components thereby resulting in the classification of the analyte and/or the determination of its concentration [12].

Chemical sensors are used to detect and monitor various gases in many different applications [28]. Environmental monitoring can be achieved by detecting the presence of toxic gases and monitoring their concentrations to create a safer and healthier environment to live in. Chemical sensors can also be used for quality control and industrial monitoring purposes specifically in industries such as food processing, beverage, perfume and other consumer products which contain chemical that need to be supervised. The constituents of the sample gas can be analyzed and determined using gas chromatograph-mass spectrometer, but most applications need systems that are smaller in size and more portable [36]. WSNs can be deployed for such applications as they possess all the required features for remote monitoring of the harmful gases without human involvement.

Popular chemical sensors used in WSN applications are metal oxide semiconductor (MOS) chemical sensors that are made up of one or more transition metal oxides such as tin oxide (SnO_2) or aluminum oxide (Al_2O_3). The first MOS sensor was devised by Taguchi and Seiyama in 1960s and it was a liquid petroleum gas detector [24]. The metal oxides are processed to form a paste. This paste is used to form a bead-type sensor. Instead, if the metal oxides are vacuum deposited onto the silica chip, thick film or thin film sensors can be made. A heating element that is made of platinum or its alloy wire is used to regulate the sensor temperature. When the sensor is exposed to gas, the metal oxide breaks the gas into charged ions. The heater heats the metal oxide to a temperature that is optimum for detection of these charged ions. The conductivity of the metal oxide changes as a result of the interaction of the gas molecules. This change is measured using a pair of electrodes. High gas concentrations produce very strong signals. Today, MOS sensors are widely used in leakage and fire detection and the sensing material has been extended using tungsten oxide (WO_3) and indium oxide (In_2O_3).

Small form-factor chemical sensors that can be used in WSNs are, for instance, sensors manufactured by ams® [44], Fig. 3.9. It is a thick film sensor that has SnO_2 deposited on two different types of substrates, namely alumina substrate and Si-micromachined substrate. This porous SnO_2 thick film is called the sensing layer and it adsorbs oxygen and water vapor present in the ambient air. Target gases are those whose detection is desired. Sensing of target gases which are reducing agents such as CO or H_2 is achieved when a reaction takes place between the adsorbed oxygen and water vapor related species. Such a reaction decreases the resistance of the sensing layer in accordance with concentration atmospheric gaseous composition of the target gas. Figure 3.9 shows the molecules of the target gas reaction on the surface of chemical sensor.

For gases that are good oxidizing agents such as NO_2 the resistance increases. The amount of change depends on the characteristics of the substrate, sensing layer and the ambient temperature. These three parameters can be varied to achieve different sensitivity towards different gases. The substrates have electrodes that facilitate the measurement of resistance of the sensing layer, allowing for even low concentrations of a large number of toxic and explosive gases can be detected and monitored. These sensors have low power consumption and can be operated in wide temperature range and humidity range. These features of the sensors along with being small in size are very essential for the wireless sensor networks applications. The relationship between the resistance of the sensing layer, R, and the concentration, C, of the gas are given by

$$R \cong K \cdot C^{\pm n}, \tag{3.5}$$

Fig. 3.9 Molecules of the target gas reaction on the surface of a chemical sensor (reproduced by permission of ams, [44])

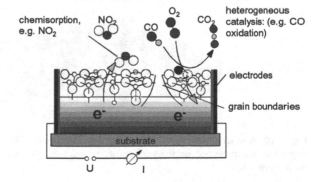

where K is a constant and n varies from 0.3 to 0.8. The positive and negative signs are to be used for the oxidizing and reducing gases.

Microcantilever (piezoelectric or piezoresistive) sensors gained in popularity with the development of micro electro-mechanical systems (MEMS) technology [25]. The sensing principle is based upon changes in the deflection induced by environmental factors in the medium where a microcantilever beam is immersed, Fig. 3.10. Bending of the cantilever induces the potential difference on opposite sides of the piezoelectric layer providing an information signal about the detected chemicals [39]. The novel sensors offer many advantages including higher sensitivities, simplified sensing systems, and lower costs.

The cantilever beam consists of the following layers from the bottom up: SiO_2, Pt, Si_3N_4, piezoelectric material (ZnO), Pt, and chemical recognition agent deposited on the top of Pt. These layers have the same geometry and dimensions but different thicknesses. In a chemical detection process, a recognition agent on the top of the cantilever reacts with the targeted chemicals in the air, soil, or solution (examples include relative humidity, mercury vapor and antibody–antigen interactions), and incudes a surface stress on the top of cantilever, resulting in a tip deflection of the cantilever beam [34]. Sensitive detection of chemicals is achieved by measuring the deflection-induced voltage generated in the piezoelectric layer.

Piezoelectric materials strain when exposed to a voltage and, conversely, electrical charge accumulates on opposing surfaces and produces a voltage when strained by an external force. This is due to the permanent dipole nature of these materials. When chemical interactions occur on the top of the cantilever, the tip displacement z caused by the differential surface stress δs can be written as [4]

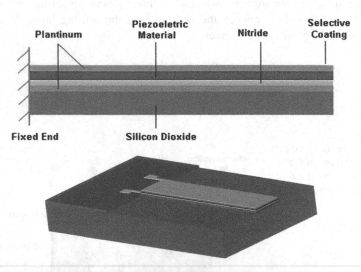

Fig. 3.10 A silicon wafer with SiO_2 layer and thin layer of Pt, then Si_3N_4 and ZnO films are deposited and patterned on the surface of the microcantilever beam. ZnO film is used as the piezoelectric layer

$$z = \frac{3(1 - v)L^2}{T^2 E} \delta s, \tag{3.6}$$

where L is the length of the cantilever, T is overall cantilever thickness, v is Poisson ratio, and E is Young's module. Assuming a thin piezoelectric layer on a thick elastic substrate and without the external force or moment [5], the relationship between the cantilever tip displacement and the corresponding voltage is given by

$$z = d_{31} \frac{3L^2 E_p}{T^2 E_e} V, \tag{3.7}$$

where V is voltage generated or applied on the piezoelectric layer, d_{31} is piezoelectric constant of the piezoelectric material, E_p and E_e are the Young's modules of elasticity for the piezoelectric and elastic materials, respectively. Note that, the thin piezoelectric layer is assumed and that the thickness of the elastic material is approximated with the thickness of the whole cantilever beam. From Eq. (3.7) one can express the induced voltage V in terms of the tip displacement z as

$$V = \frac{T^2 E_e}{3 d_{31} L^2 E_p} z. \tag{3.8}$$

Considering the Young's module E in equation as the equivalent Young's module of the whole multilayer beam and including (3.8) yields

$$V = \frac{T^2 E_e}{3 d_{31} L^2 E_p} \frac{3(1 - v)L^2}{T^2 E} \delta s \tag{3.9}$$

$$V = \frac{E_e}{d_{31} E_p} \frac{(1 - v)}{E} \delta s. \tag{3.10}$$

Equation (3.10) models cantilever beam sensor and shows the relationship between the induced voltage that can be measured and the induced surface stress caused by the chemical reaction on the top of the cantilever.

Figure 3.11 shows a wireless sensor network for chemical agent monitoring and detection. The system consists of unattended ground sensor nodes that can detect three chemicals simultaneously, namely CO, NO_2, and CH_4. Sensor nodes periodically measure gas concentrations (turning on the heather first then sampling the sensor), and if the concentration rises above pre-specified threshold value, the nodes transition into an alert mode with more frequent data sampling. Data are also sent to the base station to alert the operator.

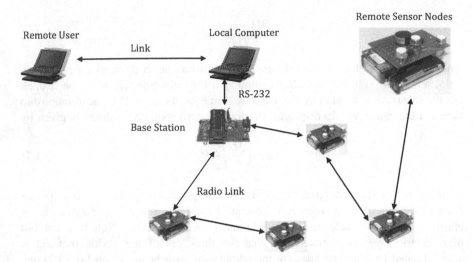

Fig. 3.11 WSN for chemical agents monitoring application

Pressure Sensors Most of the common pressure sensors convert the pressure to motion of a mechanical component of the sensor [7] such as diaphragms or combs. Bourdon tubes are pressure sensors that expand in the presence of the increased pressure, and this deflection is a function of a measured pressure. Pressure sensors can be large-scale sensors or micro-machined, MEMS-based sensors. The pressure difference causes diaphragm deflection, which is then converted into useful sensor signal. Diaphragm can have strain gauges to convert the strain into the voltage signal, or can have piezoelectric or piezoresistive transducers, or capacitive trans-ducers, etc. Figure 3.12 illustrates a cross-section of a diaphragm-based pressure sensor. The sensor can be sealed so that the reference pressure is a constant value, and the sensor then measures the absolute pressure.

The sensor diaphragm top view can take various shapes. In case of a circular diaphragm and small deflections, the diaphragm displacement is given by [7]

Fig. 3.12 Cross-section of a diaphragm-based pressure sensor

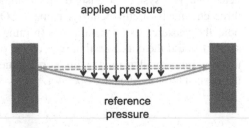

$$w(r) = \frac{Pa^4}{64D}\left[1 - \left(\frac{r}{a}\right)^2\right]^2, \tag{3.11}$$

where w is deflection, r is radial distance from the center of the circle, a is a diaphragm radius, P applied pressure and the constant D is given by

$$D = \frac{Eh^3}{12(1 - v^2)}, \tag{3.12}$$

where E is the Young's modulus, h is the thickness, and v is the Poisson's ratio of the diaphragm. From this model it can be seen that the diaphragm deflection is proportional to the applied pressure. That is, the case for small displacements relative to the size of the diaphragm.

Piezoresistive pressure sensors have piezoresistors mounted on or built in a diaphragm. For thin diaphragms and small deflections, the resistance change is linear with the applied pressure. Capacitive sensors also use diaphragm, but movement detection is based on the change of the capacitance. One capacitor plate is the diaphragm, and another plate is fixed below the diaphragm. The movement of a diaphragm is translated into the capacitance variations, which in turn is translated to the useful voltage signal using interface circuits. For a circular diaphragm with small deflections, the capacitance is given by Dunkels et al. [7]

$$C = \int\int \frac{\varepsilon}{d - w(r)} r \, dr \, d\theta, \tag{3.13}$$

where ε is the permittivity of the capacitor gap, and d is the distance between the plates. Compared with piezoresistive-based pressure sensors, capacitive-based pressure sensors offer increased sensitivity and decreased dependence on temperature variations. Modern pressure sensors have integrated electronic interfaces and sensors on the same substrate. While this is not essential for piezoresistive-based pressure sensors, it is an important design concept for capacitance-based pressure sensors due to parasitic capacitances that require on-chip signal conditioning.

Acoustic Microphone These types of sensors transform the sound waves into electronic signals that represent the sound. Microphones are widely used is cell phones, tables, and other portable devices. Most microphones are condenser microphones that consist of two plates—a diaphragm (flexible plate) that responds to the acoustic waves and a fixed, reference plate that forms a capacitor with the diaphragm. For any capacitor to work and have voltage on its plates, they must be polarized. The microphone can be polarized with an external voltage source or the material of which the plates are built can be pre-polarized as in electret microphones. Electret is a dielectric material that generates an almost permanent electric field, thus removing the need for a power supply in microphone devices. Electret microphones are a common choice because of their small size and a wide frequency range between 10 Hz and 30 kHz (humans can hear sounds in frequency range

Fig. 3.13 Electret microphone: dust cover, capsule, electret diaphragm, and amplifier module (reproduced by permission of Open Music Labs, [45])

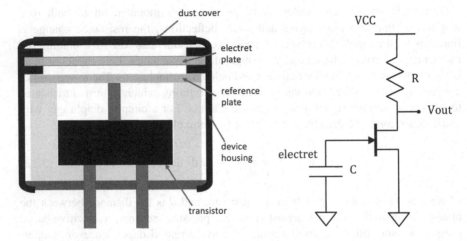

Fig. 3.14 Schematic of an electret microphone design

approximately from 20 Hz to 20 kHz, dogs up to 60 kHz, and mice up to 80 kHz). The typical electret microphone in Fig. 3.13 consists of a dust cover, capsule, electret diaphragm, and amplifier module [45].

The sound airwaves excite the diaphragm that moves back and forth, changing the distance between two plates, i.e., the electret plate and another reference plate. Such change in distance creates a voltage difference that is amplified into an output electric signal. The reference plate is connected to the gate of a JFET transistor with the amplified output measured at the transistor drain. Figure 3.14 shows the schematic design of the electret microphone.

The reference plate is connected to the JFET transistor gate, which is in a common source configuration. The electret plate and the reference plate form a capacitor, see Fig. 3.14 for an equivalent electrical model. The voltage across the capacitor is given by

$$V = \frac{Q}{C},$$

(3.14)

where Q is the charge on the capacitors plates, and C is the capacitance. The movement of the diaphragm causes a change of the capacitance:

$$C = \varepsilon_0 \frac{A}{d},$$

(3.15)

where A is the area of capacitor plates, d is distance between the plates, and ε_0 is a permittivity of the free space. Note that the voltage on the capacitor is then linearly proportional to the distance between the plates

$$V = \frac{Q}{\varepsilon_0 A} d,$$

(3.16)

meaning that the voltage on the capacitor will replicate the acoustic pressure signal. Such change in the capacitor voltage is equal to the voltage between the gate and the source, causing the change in the drain current. Voltage across the resistor R changes and is measured as an amplified equivalent of the diaphragm movement. JFET transistor is used as an amplifier of the small microphone signals.

Commonly used microphones in sensor networks applications can be classified as omnidirectional or directional. Omnidirectional microphones can detect the amplitude of sound waves, but not a direction. Directional microphones detect direction of sound waves and can achieve that by specific design or by combining two or more omnidirectional microphones with appropriate signal processing. The small form-factor directional microphones have found their applications in people hearing aids; see [30]. Figure 3.15 shows an omnidirectional and Fig. 3.16 shows an equivalent directional acoustic microphone. The omnidirectional microphone consists of the front and rear volumes, separated by the diaphragm that serves as a one plate of the sensing capacitor. As the sound waves enter the sensor chamber, the diaphragm moves back and forth and changes the capacitance of the capacitor that is formed with the electret backplate. Such capacitance changes are transferred into the output sensor signal. Directional microphones have two entry ports for the

Fig. 3.15 Omnidirectional microphone

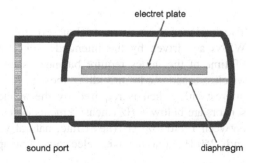

sound port

electret plate

diaphragm

Fig. 3.16 Directional
microphone

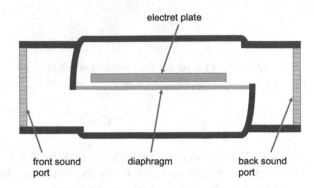

electret plate

front sound   diaphragm   back sound
port   port

sound waves. Sound waves enter the front and back of the sensor and create a pressure difference on the diaphragm (pressure-gradient microphone design). Such pressure difference creates the directional information of the microphone sensor.

The sound waves that enter the directional sensor have a slight time delay between the front port wave and the back port wave [30]. The time delay causes the phase shift between the signals in the front and back of the microphone chamber. The phase shift is angle dependent, allowing one to extract the information about the direction of the incoming sound wave. It can be shown that the diaphragm displacement is proportional to the $\cos^2\theta$ of the incoming sound signal. However, due to effective subtraction of the pressure from both sides of the membrane, the sensitivity of the directional microphone is lower than the omnidirectional one. Directional microphones can also be designed based on two directional ones, where signal subtraction is achieved during the signal-processing phase. Such methods are called dual-microphone processing and the effect of pressure difference is achieved with two omnidirectional microphones and signal subtractions and processing. The processing can be done in either analog circuitry after sound detection or in a digital form. Moreover, the directional pattern in such devices can be controlled by adding an additional programmable delay between two microphones. The adjustable delay changes the phase shift and allows the user to adjust sensitivities for the specific sound directions.

3.1.2 Microcontrollers and Microprocessors

The hardware design and a choice of microcontrollers and microprocessors for WSNs are driven by the intended application class. Limited power supply and lifetime of the nodes require balanced sampling rate of the sensors. Sensors on wireless sensor nodes are idling most of the time and measure the phenomena of interest only when is required by the application. Very low sampling rates for sensors are below 1 Hz (measurements of temperature and pressure), low rates are between 1 and 100 Hz (heart rate, natural vibrations), mid range is between 100 and 250 Hz (earthquake, electro-cardiograph measurements), high frequency sampling rates are above 1 kHz (industrial applications, audio signals), and very high sampling rates are above 1 MHz (video signals), [11]. Most designs attempt to reduce power consumption, as this presents the most important QoS factor for WSNs. Common techniques and technologies for power savings include:

- Use of subliminal processors that can operate with the supply voltage of less than the threshold voltage. The power consumption scales as a square of the DC voltage VCC. Scaling down the voltage VCC will reduce the power supply, but such reductions have their own limits below which circuit components will not be able to operate. There is a fundamental threshold voltage and subliminal processors are designed to operate at or below such voltage.
- Use of asynchronous circuits exploits event-driven nature of WSN applications. The clock is not driven when the microprocessor and other components are idling, thus reducing the power supply. A form of a handshake is needed for components to communicate.
- Power supply gating is used to switch off power supply for components that are not currently used. Transistors and their gates are used as switches that can turn off the power supply for circuit sections that are not used.
- Most of the WSNs applications are event-driven and respond only when there is a measurement of the physical phenomena that needs to be reported. However, most of the general-purpose microprocessors are not event-driven. Therefore, there is a need to design microprocessors that will be suitable for WSN-type of applications: a low-power, event-driven architecture with a large number of peripheral ports for support of large number of sensors and actuators.
- Application acceleration is provided in specific hardware architectures that accelerates common tasks and reduces energy consumption. Such design improves the WSNs performance including energy savings.

Here we provide an overview of general-purpose microprocessors that are most commonly used in WSN platforms as well as few special purpose microprocessors [26] that are designed with specific WSN applications in mind including event-driven microprocessors, subliminal microprocessors, asynchronous circuits, and others [11].

General Purpose Microprocessors Those are microprocessors that are commonly used in embedded systems applications as well as WSNs. They offer flexibility in programming, have a variety of peripheral devices, can easily be interfaced with numerous sensors and actuators, and satisfy basic QoS requirements for WSNs. Such microprocessors that are commonly used in WSNs platforms are Texas Instruments MSP430 and Atmel ATMega 128L. Both microprocessors are low-power, general-purpose computing and control devices that are not designed for event-driven systems such as WSNs, but can be used when accompanied with appropriate operating system such as TinyOS (an event-driven operating system for WSNs).

TI MSP430 [41] is a 16-bit, RISC-based, microprocessor designed specifically for ultra-low-power applications. The microprocessor has variety of low-power modes with a large number of peripheral devices and a smart clock system. The clock system allows one to disable clocks and oscillators currently not in use, thus reducing the power consumption in idling modes. The CPU runs at clock speed ranging from 8 to 25 MHz, supports up to 512 kb of Flash memory and up to 64 kb of RAM memory. The microprocessor has a wide variety of intelligent peripherals that have an option to function independently of CPU. They include A/D converters, secure digital (SD) cards, D/A converters, operational amplifiers, comparators, DMA, USB, UART, and various timers. The system is connected through common memory address bus (MAB) and memory data bus (MDB), see Fig. 3.17.

The CPU is based on a RISC architecture that supports hybrid clock system with various clocks and oscillators. Such design is suitable for low-power applications with numerous peripheral devices by enabling required clocks only when they are needed. The clock system includes main clock (MCLK) where CPU is the source and is driven by the internal oscillator up to MHz or with external crystal, auxiliary clock (ACLK) that is used for peripheral modules and is driven by the internal low-power oscillator or external crystal, and sub-main clock (SMCLK) that is a clock for faster individual peripheral modules that may be driven by the internal oscillator up to 25 MHz or with external crystal [41]. The microprocessor can wake up almost instantly by internal oscillator in <1 μs. The device has ferroelectric random access memory (FRAM) that combines the capabilities of RAM with that of a flash memory. FRAM is a non-volatile RAM that is lower power, faster, and supports larger a number of erase cycles than flash technology.

TI CC43 is a wireless networking microprocessor device from Texas Instruments. Trends are to integrate as much functionality as possible in the same chip to reduce space, power, and cost of the final product. In this case combining RF module with the TI MSP430 microprocessor resulted in a low-cost, low-power device that is suitable for WSN applications. TI CC430 includes an RF transceiver

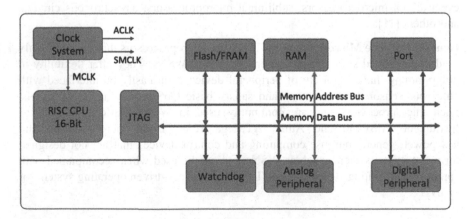

Fig. 3.17 Architecture of Texas Instruments MSP430 [41]

for frequencies below 1 GHz, flash memory, RAM, timers, A/D and D/A converters, DMA, LCD support, and more. The RF transceiver is based on CC1100 module, now integrated with the microprocessors for a system-on-a-chip for wireless sensing and monitoring applications. It supports Industrial, Scientific and Medical (ISM) bands at 315, 433, 868, 915 MHz and can be programmed to other frequencies in that range. The transceiver has an integrated baseband modem with programmable data rate and modulation techniques, [41]. For more details on the RF transceiver, please see the Sect. 3.1.3.

Atmel ATmega 128L is a high-performance, low-power Atmel 8-bit AVR type RISC microcontroller that has 128 kb of programmable Flash memory, 4 KB SRAM, a 4 kb EEPROM, A/D and D/A converters, and a JTAG interface for on-chip debugging [40] with a throughput of 16 MIPS at 16 MHz. An architecture of the Atmel's AVR microcontroller is given in Fig. 3.18 applying a Harvard architecture with separate data and program memory and bus.

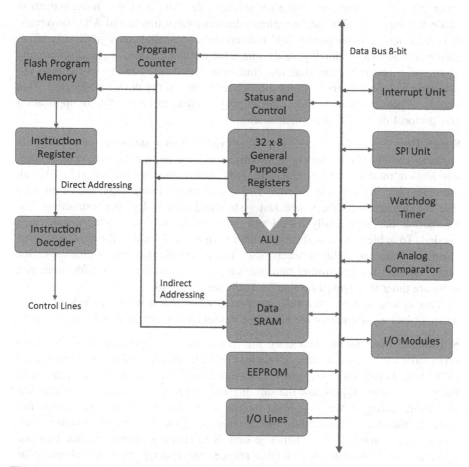

Fig. 3.18 Architecture of Atmel AVR microcontroller [40]

The microcontroller has 32 general-purpose registers that are accessible by the arithmetic logic unit. By executing instructions in a single clock cycle, the device achieves throughputs approaching 1 MIPS per MHz, balancing power consumption and processing speed. The operating voltage is between 2.7 and 5.5 V. The microcontroller includes a large variety of peripheral devices including two timers, two counters, two pulse-width modulation (PWM) channels, six PWM channels with programmable resolution, comparators, A/D and D/A convertors, dual serial USART, SPI serial interface, and some special microcontroller features including external and internal interrupts, clocks that can be programmed, programmable I/O lines, and more.

The microcontroller has several power savings modes that are suitable for WSN applications. They include the idle mode where the CPU stops, while peripheral devices including SRAM, timers, counters, and others continue to function; power-down mode where the register content is preserved but the rest of the chip is disabled until the next interrupt; power-save mode where timers is running and other parts of the microcontroller are inactive; the A/D converter noise reduction mode that stops the CPU and peripheral devices except timers and A/D converters, thus reducing the noise during A/D conversion process that can significantly affect conversion accuracy; standby mode where the oscillator is running and other parts of the microcontroller are sleeping; and extended standby mode where both the oscillator and asynchronous timer run, while the rest of the device is sleeping. This variety of power savings modes allows programmers to be creative in application and protocol designs and implementation.

Smart Dust Smart dust is one of the first event-driven systems designed for WSN applications. The smart dust system consists of sensors, a solar cell, optical communication module, and a Harvard, RISC-based microcontroller [11, 33], all integrated into 1 mm-scale volume. The system uses a low-energy microcontroller that is integrated with the smart dust system and uses <12 pJ per instruction. The microcontroller is normally halted and is waken up only to execute a task when needed. To achieve this, a slow oscillator is integrated for a real-time clock operation and supports multiple timers. Two timers sample two sensor channels. One timer supports the transmitter and one supports the receiver. The fifth timer is a software timer that wakes up the datapath, see Fig. 3.19.

This system was the first fully integrated sensor node where all WSN components fit into a millimeter-size hardware module justifying its name—smart dust.

Subliminal Processors Developed for sensor networks applications, such microprocessors are extremely energy efficient as they operate in sub-threshold regime [37]. Authors call such microprocessors Subliminal Processors. The system scales the power voltage V_{DD} below the sub-threshold level, to save energy consumption. However, scaling down V_{DD} level has its limits, i.e., when leakage energy and dynamic energy are comparable since leakage energy increases with reducing of the power supply voltage. The challenge here is to create a microprocessor that will operate in sub-threshold regime that reduces the leakage current. Sub-threshold processors have simple control complexity because they have compact circuits with

Fig. 3.19 Block diagram architecture of a smart dust millimeter-scale system consisting of sensors, a solar cell, optical communication, power with battery, and electronics with digital signal processing and control

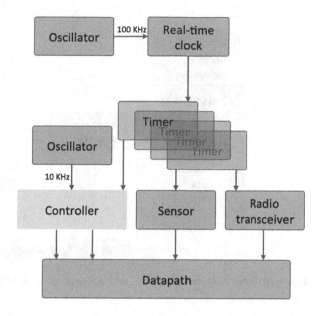

high activity rate and a low leakage/dynamic current ratio. In [37], a sub-threshold microprocessor is designed resulting in 360 mV operating voltage that consumes 2.6 pJ per instruction and operates at 833 kHz. The microprocessor consists of two pipeline stages, a memory for register file, pointer file, instruction memory, data memory, an 8-bit wide arithmetic logic unit, and 32-bit accumulator, see Fig. 3.20.

The memory was implemented using a custom multiplex-based array structure. Register file, pointer file, instruction memory and data memory are implemented in one SRAM. The microprocessor was implemented in 130 nm CMOS technology. The microprocessor is a result of optimization between instruction set size and control logic complexity. For the optimal operating voltage, a simple 32/16/8-bit single operand instruction set architecture was selected with two register banks, one 4-entry 32-bit integer register file and one 4-entry 16-bit pointer register file. The proposed architecture can address up to 64 KB of memory. Readers are referred to [37] for more details on the microprocessor design.

Sensor Networks Asynchronous Processor (SNAP) This microprocessor is developed specifically for sensor network applications and is based on asynchronous RISC architecture [8]. It is an event-driven processor that allows efficient power control since the microprocessor is in an idle (low-power) state most of the time. Since low duty cycle applications are typical for WSN monitoring applications, this type of processor is then very suitable for WSN systems. When event of interest occurs, the SNAP transitions to an active state in few 10s of nanoseconds.

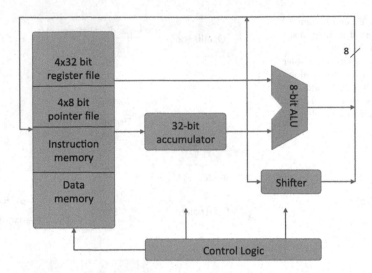

Fig. 3.20 Subliminal microprocessor architecture proposed in [37]

An asynchronous design of the microprocessor means simplified hardware that does not require an operating system to run on the microprocessor and thus reduces the microprocessor instruction set. WSN nodes with general-purpose microprocessors support event-driven applications using software such as TinyOS. The software handles incoming interrupts that process sensor information as a result of measured event. On the other hand, SNAP implements an event-driven operation through the specific hardware design with an event queue and event coprocessor, thus reducing any software overhead. Moreover, the interface circuit between the microprocessor and the radio and sensors is simplified as well. As a result, the microprocessor consumes very low energy per operation.

As an asynchronous, event-driven design, SNAP does not use the clock. The circuits use handshaking protocol to implement synchronization between various hardware modules. Since SNAP has low-power sleep mode, long battery lifetime compared with other WSN designs, and very low wake up time in order of nanoseconds, it is optimized for low-power operation in a sensor network.

The microprocessor architecture is shown in Fig. 3.21 and consists of three major components: processor core that includes event queue, instruction fetch, decode, execution units, register file, message queue, instruction memory and data memory; timer coprocessor; and message coprocessor [8].

The microprocessors functions using instruction tokens where each computational block waits for the token. The instruction tokens travel through the system pipelines and computational blocks. After the necessary boot code, the microprocessor waits for the event token. Such token can arrive either from the message coprocessor or timer coprocessor. In case of reading the sensor data, it is a message coprocessor that sends an event token. Arrived token will be fetched and decoded, and microprocessor will execute set of instructions that correspond to that specific

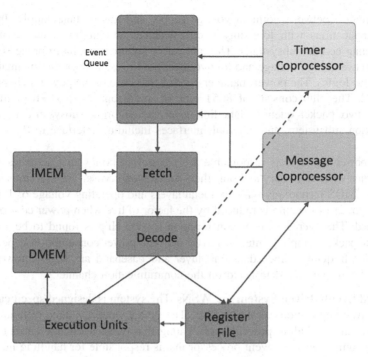

Fig. 3.21 Architecture of the sensor network asynchronous processor [8] that consists of the processor core, timer coprocessor, and message coprocessor

event. If the even queue is empty, the microprocessor will transition to an asleep mode. There are two ways to communicate with external sensors—passive when sensors activate interrupt pin that is connected to the message coprocessor, which in turns initiates a token into core's event queue; or active when the core sends a command to the message coprocessor to read data from the sensor.

SNAP microprocessor was compared with Atmel Mega128L microcontroller running TinyOS. Throughput is approximately 28 MIPS at 0.6 V compared with 4 MIPS at 3 V on Atmel's microcontroller. The transition from idle to active mode is approximately 2.5 ns at 1.8 V, while Atmel microcontroller requires more than 4 ms. SNAP requires <300 pJ per instruction (at 0.6 V energy consumption is even lower at 23 pJ per instruction), while Atmel microcontroller requires five times more. For low duty cycles applications, SNAP uses several orders of magnitude less energy than general-purpose microcontroller such as Atmel Mega128L.

Charm This is a digital protocol processor that implements a radio stack specific for WSN applications [26]. The chip is divided into subsystems that correspond to OSI reference model and includes the application, network, data link layer, radio baseband, neighborhood management, and localization subsystem with support of 64 kb of RAM. Moreover, the hardware implementation has power management capabilities where each subsystem has independent power control with switching power control between nominal voltage supply and low voltage supply. In most WSN applications with low duty cycles, leakage current limits the savings of deactivating certain subsystems. The innovative solution in Charm processor is to switch between high voltage and low voltage (the lowest voltage that maintains the state in the logic). The power manager controls the state of the power switches, see Fig. 3.22. The chip consists of 8051 type of microcontroller, 64 kb of memory (RAM), two packet queues, data link layer, neighboring subsystem, baseband, localization subsystem, and external interfaces including interface to the external radio.

The power manager is responsible for scheduling policy that activates certain subsystems when necessary and puts them to sleep otherwise. The system is built in 130 nm CMOS technology with six metal layers and operating voltage of 1 V. The leakage power of the chip is reduced by the factor of five when power control logic is enabled. The average power consumption for the chip is found to be 132 μW when the packet sampling interval is 100 ms. The power consumption depends on the packet frequency since data link layer and baseband are largest power consumers due to periodic data listed on the communication channel.

Harvard Event-Driven System for WSNs The system is designed specifically for event-driven applications in WSNs [11]. The system employs three techniques for power savings including power supply gating that reduces leakage current of the sleeping components, an event processor that is responsible for handling incoming interrupts, and hardware acceleration for tasks related to WSN applications. The structure of the system is given in Fig. 3.23. The system has 4 KB of memory and runs on power supply between 0.5 and 1.2 V with 12.5 MHz minimum clock frequency. Leakage current savings that are achieved through V_{DD} gating are two orders of magnitude compared with standard microcontrollers.

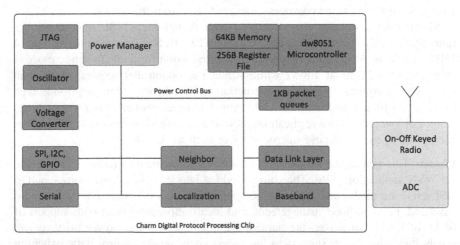

Fig. 3.22 Architecture of the Charm protocol processor [26]

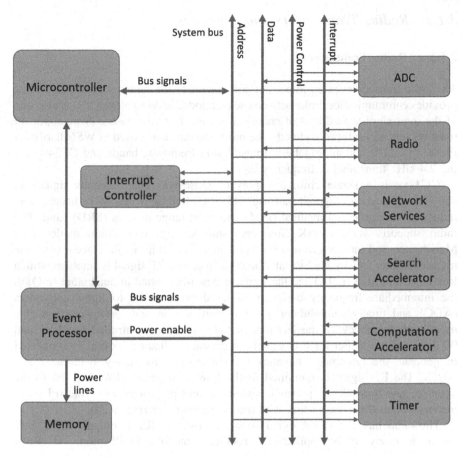

Fig. 3.23 Architecture of the Harvard event-driven system for wireless sensor networks applications [11]

The system in Fig. 3.23 consists of master components that can control the bus lines and slave components that are specific for WSN applications [11]. The bus lines have data, address, power control, and interrupt lines. The master devices are microcontroller and event processor that handles interrupts. Slave devices include AD converters, radio, network services accelerator, search accelerator, computation accelerator, and timer. Those slave devices allow for efficient power management where they can be turned off or on based on need, all while the main microcontroller is in sleep mode. The microcontroller is a general-purpose CPU. The event processor handles interrupts by executing interrupt service routines. The event processor communicates with the slave components. The slave components have four 16-bit timers that support sensor network applications.

3.1.3   Radios Transceivers and Antennas

3.1.3.1   Radio Transceivers

The radio transceiver is an important component of the wireless sensor node that provides communication links between sensor nodes and/or gateways. It is also one of the most significant drains of energy in WSNs. For this reason, this component must be very carefully considered. The most popular radios used in WSN hardware include CC1101 for sub-gigahertz transmission frequency bands and CC2400 for the 2.4 GHz transmission frequency band.

CC11xx is a single chip, low power, transceiver that transmits signals in sub-GHz frequency range; it can use 315, 433, 868, and 915 MHz license-free, industrial, scientific and medical (ISM) and short range devices (SRD) band. The radio supports various FSK (frequency-shift keying) modulation modes with Manchester and non-return-to-zero encoding. The chip is interfaced with the microcontroller usually via SPI interface. The received RF signal is amplified with a low-noise amplifier (LNA) and the signal is down-converted in quadrature (I, Q) to the intermediate frequency before it passes through analog-to-digital converters (ADC), and finally demodulator. On transmitting side, the system consists of a modulator, frequency synthesizer that includes a voltage-controlled oscillator and 90 degree phase shifter for I, Q signals. A crystal oscillator is externally connected to generate the reference frequency for the on-chip frequency synthesizer and ADCs. The RF signal is amplified in the Power Amplifier (PA) and fed to the antenna. The transmitting power is adjustable and programmable. The SPI serial interface is used for connection with the microcontroller (Fig. 3.24).

The radio has a received signal strength indicator (RSSI) output that is very useful in many WSN applications, routing protocols, localization, etc. RSSI

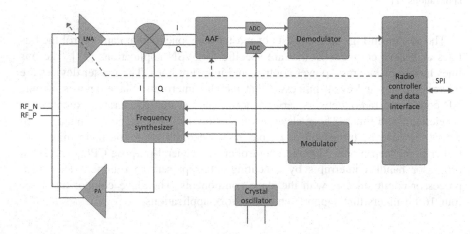

Fig. 3.24 Block diagram of the Texas Instruments CC1101 radio [41]

measures of strength of the received signal and provides approximate estimate of the distance between the transmitting and the receiving sensor node.

CC2420 is also a low-power, single chip radio that operates in 2.4 GHz band and supports 802.15.4 and ZigBee protocols. A key feature is its ability to easily implement spread spectrum techniques such as Direct Sequence Spread Spectrum (DSSS). This protocol, along with offset quadrature phase-shift keying (O-QPSK) modulation allows many devices to share the same frequency channels efficiently and without interference. The chip has support for packet handling, data buffering, encryption, authentication, and packet timing indication. It also uses SPI serial interface for communication with the microcontroller and has RSSI support. The transmitting power is programmable.

The received signal strength (RSS) model is given by Mao et al. [19]

$$P_{ij}(\text{dBm}) = P_0(\text{dBm}) - 10 \cdot \alpha \cdot \log(d_{ij}/d_0) \tag{3.17}$$

where α is a path loss exponent (PLE), $P_{ij}(\text{dBm})$ is power received at node j from node i in dB milliwatts and $P_0(\text{dBm})$ is a reference power received at some distance d_0, and d_{ij} is a distance between nodes i and j. The PLE depends on the medium of electromagnetic signal propagation (including obstacles). Most radio chip manufacturers provide a more specific RSS model for various transmitting node (radio) power settings that relate the signal strength, distance from the transmitting radio where such RSS is measured and the PLE.

AT86RF231 is Atmel's popular radio chip that also supports EEE 802.15.4 and ZigBee protocols using 2.4 GHz. The radio consumes ultra-low power and also has a programmable output power. It is a highly integrated solution for WSNs since there are no any external components besides the oscillator crystal, capacitors, and antenna. Integration with a microcontroller is through the serial SPI interface.

3.1.3.2 Antennas in WSNs

The main function of antenna is to convert electric signals into radio waves and vice versa. There are different kind of antennas used with radio chips and other electronics components. Some of the commonly used short-range antennas, applicable for WSNs, are monopole, dipole, helical, loop, and PCB antennas.

Antenna design influences the transmission range and connectivity in WSNs. Since WSNs nodes are usually very small form-factor, antenna size is also restricted. This creates some constraints and trade-offs in antennas for WSNs. For a very small size of sensor nodes and antennas, the common solution is a printed circuit board (PCB) antenna—an antenna that is integrated with the WSN node circuit. If a longer transmission range is required, external resonant antennas are used.

Common factors that determine antenna selection in the design process include: antenna gain—antenna's ability to radiate power in a certain direction when connected to the source; antenna directivity—radiation pattern of electromagnetic waves, with *directional antennas* having a higher gain in certain directions and *omnidirectional antennas* having an uniform radiation pattern in every direction; antenna efficiency—a ratio of power radiated from the antenna and the power dissipated in the antenna (heat); antenna Q-factor—a ratio of antenna reactance and antenna resistance, a parameter that describes antenna resonator characteristics with a preferred low Q-factor in antenna design as this means larger bandwidth and easier matching and tuning; antenna bandwidth—a range of frequencies where antenna can operate with a certain efficiency; antenna polarization—the polarization of the wave radiated by the antenna, where polarization of the wave is considered an orientation of the electric field of the corresponding electromagnetic wave; for more detailed antenna theory see [9].

For an efficient transfer of energy, the impedance of the radio, of the antenna, and of the transmission cable connecting them must be equal. Radio components and an open-air transmission medium typically have 50 Ω impedance, therefore, requiring antenna impedance to be close to that value. If that is not the case, and to avoid transmission losses, an impedance matching circuit is needed to provide an impedance match.

Antennas are resonant structures, i.e., they resonate at the frequency of operation. For proper resonating of an antenna, the size of antenna is an integer multiple or a fraction of the signal wavelength. In case of a monopole antenna, see Fig. 3.25, optimum antenna length is *one quarter of the signal wavelength.*

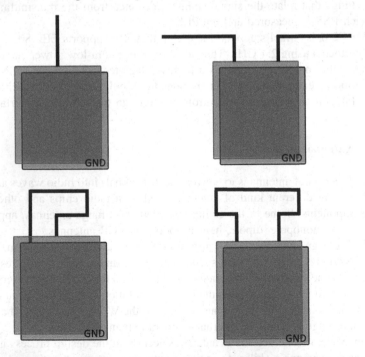

Fig. 3.25 Top and ground layer of a PCB sensor node with different antenna designs: monopole (*upper left*), dipole (*upper right*), single-ended loop (*lower left*), and differential loop (*lower right*) PCB antennas [9]

PCB antennas are recommended for small form-factor WSN modules where external antennas are not feasible due to size or physical limitations. The RF circuits and hardware modules are required to have a ground plane within the PCB. The ground plane should not cover the antenna area to avoid interference with the antenna and reduction of antenna efficiency.

Antenna's feed signal ports and ground connections are shown in Fig. 3.25. The monopole antenna is fed at one port and the antenna is away of the ground plane. Monopole antennas are relatively low-cost and easy to install compared to other antennas. They should be matched to 50 Ω impendence and might require a balun module (a hardware component that converts an unbalanced signal measured against the ground only into a balanced signal where two signals working against each other and ground is irrelevant, and vice versa) if the radio chip expects a differential output. The dipole antenna has differential port and this structure might also require a balun. The dipole antenna should also stay away from the ground plane. The loop antenna can have one feeding port (the other port is grounded) or two (differential feed).

In terms of *polarization and radiation patterns*, antennas needs to be designed and positioned such that there is an optimal received signal strength at the receiving sensor node. In WSN applications, sensor nodes usually establish communications in the horizontal plane. The loop antenna will have omnidirectional radiation pattern if the loop antenna is in the horizontal plane (its null is normal to the loop). The monopole and dipole antennas have radiation pattern in the plane normal to the antennas' axes and null on the antennas' axes, and therefore, should be placed vertically.

External antennas are often used when the communication range is the main design requirement such as in the base stations [9]. An external monopole, quarter wave antenna has an impedance of 37 Ω and can easily be matched to 50 Ω. Such antennas have omnidirectional radiation pattern in the horizontal plane. The external monopole antenna is the most common solution in WSNs when the ground plane is present. An external dipole antenna has an impedance of 73 Ω and requires a differential feed connection.

3.2 Layered Network Architecture

WSN nodes are typically low-power modules, with limited CPU and memory, and the application requirements and network architecture of WSNs differs greatly from that of traditional computer networks. Here we describe the major features of the WSN protocol stack, highlighting the design requirements and performance attributes of the different layers. Figure 3.26 shows the layers of the protocol stack, with Table 3.1 summarizing the functions of the different layers. Note that roles such as power, mobility and task management have to be fully engrained in the operation mechanisms of each of the five layers, in order for the network to be able to cope with the earlier mentioned WSN limitations. These layers control power usage (e.g.,

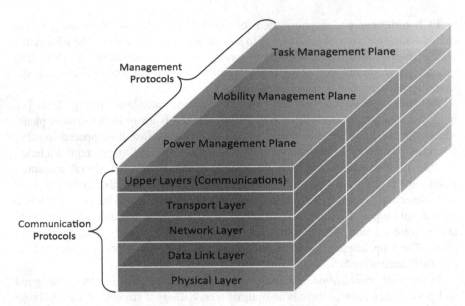

Fig. 3.26 The wireless sensor network protocol stack: each communication layer extends into power, mobility, and task management planes

Table 3.1 Wireless sensor network communication protocol stack [27]

Upper layers	In-network applications, including application processing, data aggregation, external querying, query processing, and external database
Layer 4	Transport, including data dissemination and accumulation, caching, and storage
Layer 3	Network layer: routing, topology management and routing
Layer 2	Link layer: channel sharing (MAC), timing, and locality
Layer 1	Physical medium: communication channel, sensing, actuation, and signal processing

switching on/off of radio), movement, and task sharing among the different nodes on the WSN.

3.2.1 Physical Layer

The main task of the physical layer is modulation and demodulation of data. Other roles include data encryption, frequency selection and signal detection among others [2, 14]. The two major design requirements of a WSN physical layer pertain to minimization of both the cost and power utilization of the sensor. Since sensors will typically be redundantly deployed in large numbers over a sensor field, a low-cost per unit is key in ensuring feasibility of WSN applications. It is thus important that the sensor's physical components, i.e., the chips, radios and external parts be as cheap as possible. As for the power consumption, the sensors should have low transmission power, and must be designed to run a low duty cycle, such that the battery is discharged in pulses (by activating the major power dissipating components such as transceivers in bursts) in order to minimize average power consumption. A battery undergoing pulsed discharge generally has lower mean power consumption than one undergoing constant discharge, as the pulsed approach tends to have a form of *charge recovery* effect.

3.2.2 Link Layer

Nodes in wireless sensor networks use a unique channel to communicate with each other. At any given time only a single node can transmit data through this channel. Therefore, when multiple nodes in the wireless sensor networks want to share the channel, it becomes necessary to establish a protocol named media access control (MAC) protocol. MAC protocol is designed to control the access to the shared wireless medium in such a way that the performance of the underlying application is not hampered. Below, some vital design considerations of the various MAC protocols are described.

Energy Efficiency Tiny batteries with very small capacity are typically used to power the sensor nodes. Power conservation plays a vital role in prolonging the life span of nodes in WSNs. Consequently energy efficiency has been given the most emphasis in designing the MAC protocols for WSNs. There are several causes of potential power consumption inefficiency. First among these is packet collision. Collision occurs when two or more sensor nodes attempt to transmit data over the channel simultaneously. Another vital source of energy consumption is idle listening. Idle listening refers to listening for a traffic that is not sent. Overhearing is another source of power wastage, and occurs when sensor nodes receive packets destined for other nodes. Lastly, frequent switching between various operational modes also reduces energy efficiency significantly. MAC protocols are generally designed to consider such energy restrictions.

Delay Delay refers to the time required by a data packet in the MAC layer before successful transmission over the channel. MAC protocols should provide guaranteed delay bound to ensure the quality of service (QoS) required by the applications.

Throughput Throughput refers to the rate of data packet servicing by the system. The design of MAC protocols aims to maximize the channel throughput, whereas minimizing delay ensuring performance satisfaction of the underlying applications.

Robustness Robustness refers to how much insensitive the protocol can be to errors generated by the failures of communication links and nodes of the WSNs. A good MAC protocol should have built-in mechanisms to overcome errors.

Scalability Scalability means an ability of a WSN to adapt with increase of the size of the network without degradation of performance. The size of WSNs varies greatly in terms of number of nodes, ranging from hundreds to millions.

Stability Stability refers to as an ability of the wireless sensor network to handle fluctuations of the traffic load over a time period.

3.2.3   Medium Access Protocols in WSNs

There are three primary strategies to access a shared medium: fixed assignment, demand assignment, and random assignment.

3.2.3.1   Fixed Assignment Protocols

Fixed assignment strategy assigns a certain amount of the channel resources to each sensor node. Sensor nodes can use this assigned resources exclusively. Frequency division multiple access (FDMA), time division multiple access (TDMA), and code division multiple access (CDMA) belong to this category. We discuss here in more details each of these protocols.

FDMA The radio spectrum is shared among the sensor nodes. The available channel bandwidth is divided into regions named sub-channels. Different carrier frequencies of the radio spectrum are then assigned to different nodes to transmit data exclusively. This method requires that the communicating nodes be frequency-synchronized to avoid a frequency overlapping between neighboring communication channels.

TDMA In TDMA, channel bandwidth is shared among the communicating nodes without being divided into sub channels. Radio frequency is divided into certain number of time slots and then a unique time slot is assigned to each communicating node. Nodes are then given the chance to transmit and receive packets in a round robin fashion in their respective time slots.

CDMA Spread spectrum (SS) technique is used in a CDMA method where multiple communicating nodes can transmit data concurrently. Spread spectrum is a radio frequency modulation technique where the radio energy is spread over a larger bandwidth than that required for the data transfer rate. Major spread spectrum based systems use either direct sequence spread spectrum (DSSS) or frequency hopping spread spectrum (FHSS).

3.2.3.2   Demand Assignment Protocols

Demand assignment protocols conceptually divide the nodes into two types: idle nodes that are ignored and ready to transmit and nodes that are assigned the channel for a certain amount of time to transmit data packets. This protocol typically uses a mechanism to determine which of the contending nodes should access the channel in a specific time. The demand assignment protocols can be further classified into two classes: centralized and distributed. Polling method is a representative of centralized protocol, whereas token and reservation based methods use a distributed protocol.

Polling The central control device maintains a specific order to query each node about whether it is ready to transmit data packets. If the polled node is ready to transmit the data packets, it replies positively. Getting positive reply, the controller assigns the channel to the ready node so that it can access the channel exclusively to transmit data packets. If the polled node does not have any data packet ready to transmit, it replies negatively to the controller's query. Getting this negative reply, the controller proceeds forward to query the next node in order.

Reservation In reservation based method, some small time slots are used to carry reservation messages. These time slots are called mini slots since these reservation messages are normally smaller than data packets. When a sensor node is ready to transmit data packets, it sends a request for a data slot via a reservation message in the mini slot. After receiving requests, the master node prepares a schedule for transmission of data packets and then announces the schedule to the slave nodes.

3.2.3.3   Random Assignment Protocols

In Random Assignment Protocols, ready nodes need to fight for taking the access of the channel to transmit data packets since there is no predetermined schedule or order of the nodes for data transmission. Consequently, collision occurs when more than one node attempts to transmit data packets concurrently. To resolve this problem, random assignment protocols include a collision detection method and another method to retransmit the colliding packets. ALOHA, carrier sense multiple access (CSMA), carrier sense multiple access with collision detection (CSMA/CD), and carrier sense multiple access with collision avoidance (CSMA/CA) belong to the category of random assignment protocols.

MAC protocols can also be divided into schedule-based and contention-based protocols.

3.2.3.4 Schedule-Based Protocols

At any given time only one sensor node can access the channel. Schedule-based protocols prepare a schedule to determine which sensor node can access the channel in a certain time. Generally, some modified versions of TDMA are used in schedule-based protocols. Following the order defined by the schedule, sensor nodes alternate between two modes of operation: active mode and sleep mode. In active mode, sensor nodes use the assigned slots for data transmission. After using the time slots assigned to the sensor nodes, they go to sleep mode. In sleep mode the sensor nodes turn the radio transceiver off to conserve battery power. Some protocols belonging to this category are described below.

(a) *SMACS*: Self-organizing medium access control for sensor nets (SMACS) assumes that radio spectrum is divided into many sub-channels (frequency channel) or that many CDMA codes are available. SMACS also divides its time into super frames. Super frames are of fixed length and time synchronized with neighbors. Super frames also need to be divided into smaller time slots. SMACS execute a neighbor detection algorithm to find its neighbors and then set up exclusive links to them. This link is unidirectional and has a TDMA slot in either endpoint. For bidirectional transmission, both nodes should have receive and transmit slot. To avoid collisions, SMACS ensures that times slots of different links belonging to same node do not overlap. SMACS also pick randomly one frequency channel/CDMA code and allocates to each different link. When links are setup successfully the sensor nodes are ready to transmit/receive data frames. The nodes become active once per super frame following the schedule. In the receive/transmit time slots the sensor node tunes to its frequency channel/CDMA code.

(b) *Bluetooth*: Uses one universal short-range radio channel to connect several electronic devices like mobile phones, tablets, laptops, netbooks, etc. A small set of such devices, sharing a single channel, is referred to a piconet. The piconet consists of at most eight devices where one device should act as a master and others are referred to slaves. Based on the master's Bluetooth device address, every piconet receives an exclusive frequency hopping pattern where the radio channel is subdivided into slots of 625 ms. A number of piconets interconnects using bridges and form a scatternet. The master device of the piconet assigns a three bits address to its slaves. The Time Division Duplex (TDD) protocol is used to determine who is going to access the channel in order. The master device of the piconet assigns time slots to its slaves using polling method. Finally, Bluetooth defines four operational modes: active, sniff, hold, and park to ensure energy efficiency.

(c) *LEACH*: Low energy adaptive clustering hierarchy (LEACH) is a hierarchical method that makes clusters from the sensor nodes. There will be a head node in each cluster and TDMA is used for the communication between the head and other nodes in the cluster. The main role of the cluster head is to receive messages from its cluster nodes, aggregate them and then forward to the base station. To avoid collisions, the cluster head prepares a TDMA schedule and then sends it to other cluster nodes before data transmission. The cluster node follows this schedule to become active during its assigned time slots and remain turned off at other times. LEACH uses CSMA and fixed spreading code for data transmission between cluster heads and base stations. When a cluster head is ready to transmit data to the base station, it senses the carrier to see whether it is busy (i.e., other cluster head is also sending data packet to this base station) or not. If the cluster head finds the carrier idle it will send the data following the CSMA protocol, otherwise it will wait until the carrier becomes idle.

3.2.3.5 Contention-Based Protocols

Contention-based MAC layer protocols do not prepare any schedule for the nodes to access the channel; rather they use on-demand access. When several nodes attempt to access the channel simultaneously, collision occurs. Nodes whose transmissions are involved in a collision wait for a random amount of time and then try again to access the channel. As traditional MAC layer protocols are not well suited for sensor networking, several new mechanisms such as collision avoidance, request-to-send (RTS) and clear-to-send (CTS) are being used to enrich these traditional protocols. These mechanisms have improved the performance as well as made the protocols more robust to some common and dangerous problems of sensor networks. For example, hidden terminal problem has been given special consideration to these mechanisms. Below, some contention-based protocols are described briefly.

(a) *PAMAS*: A characteristic of the power aware multi-access protocol with signaling (PAMAS) is that it uses a separate signal link to avoid overhearing, which is a big source of power waste. The protocol attaches a busy tone with RTS and CTS packets and thus allows currently inactive nodes to get turned off to conserve energy. Although this protocol considers overhearing to conserve energy, it does not provide any efficient method to lessen power waste caused by idle listening.

(b) *STEM*: The sparse topology and energy management (STEM) protocol emphasize an energy efficiency for which it uses two radio channels: a data radio channel and a wake up radio channel. In STEM, a sensor node keeps the data radio channel turned off until its ready to communicate with other nodes. When the node becomes ready to send data, it starts to send a signal using the wake up channel. This wake up signal can be thought of as paging signal. This signaling will be continued until all neighbors have been paged. When it sees all its neighbors are paged, it starts to send data using data radio channel.

(c) *IEEE 802.11*: An important feature of these protocols is the use of RTS and CTS packets for avoiding overhearing. They also prevent idle listening by ensuring synchronization between neighbors. These protocols maintain low duty cycles to reduce power waste. To reduce packet latency, these protocols ensure that the sensor node ready to send packets waits a certain amount of time before the receiving node becomes wake up. These protocols also maintain some variations from the original protocols in the level and the way of fairness among sensor nodes.

(d) *T-MAC*: The timeout mac (T-MAC) is intended for those applications that are less sensitive to packet latency and also have less number of packets to send per unit time. T-MAC handles the collisions using RTS, CTS, and acknowledgment packets. T-MAC has a special criterion that it can adapt to traffic load variation. This protocol also applies adaptive duty cycle and thus ensures energy efficiency. T-MAC takes special watch to lessen idle listening. It sends all messages in a burst. These bursts are of variable length. Nodes go to sleep mode between bursts. From the channel traffic load, T-MAC can dynamically calculate the optimal duration of active time for the nodes. A node listens and transmits only when it is in active mode. Thus, sensor nodes alternate between active and sleep modes to reduce power waste.

(e) *B-MAC*: The Berkeley media access control (B-MAC) is an energy efficient protocol. It applies clear channel assessment (CCA) for channel arbitration. To achieve reliability it uses acknowledgement message. To ensure energy efficiency it reduces duty cycle and idle listening. It reduces duty cycle through applying adaptive preamble sampling method. Each node alternates between active and sleep mode. When a node turns to active mode, it looks for activity. If the node finds any activity, it keeps the transceiver ready to receive data. When data reception is done, it goes back to sleep mode. It waits to receive data until a timeout. If no packet is received within the timeout, it automatically goes to sleep mode.

3.2.4 Network Layer

This layer defines the routing protocols used by WSNs. An in-depth discussion of WSN routing protocols is presented in the Chap. 2.

3.2.5 Transport Layer

In a typical WSN application, large numbers of sensors send traffic to a common base station. As traffic arrives close to the base station, there is every likelihood of a monotonically increasing load, which implies that during overload, congestion is very likely to bite hardest at the links closest to the base station (see Fig. 3.27 for traffic convergence illustration).

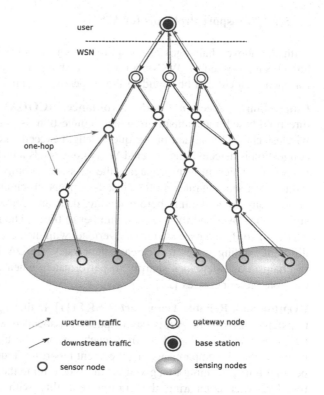

Fig. 3.27 Traffic convergence: the amount of traffic increases from the sensing nodes to the base station resulting in higher likelihood of congestion in upstream links

In some cases congestion could even begin far upstream, such that links far away from the base station also see overloaded queues. The result in both cases is that data packets get dropped, prompting retransmissions, and causing significant energy loss and transmission delays. Because of the well-known energy limitations of WSNs and the strict demands on timeliness of data delivery for most WSN applications, WSNs require transport layer protocols that have been designed to counter these problems as much as possible.

Transmission control protocol (TCP) and user datagram protocol (UDP), the Internet's predominant transport protocols, are unsuitable for several reasons. First, UDP does not have congestion control and flow control mechanisms, and thus can neither recover from losses, nor take steps to prevent loss during overload. TCP on the other hand suffers from the overhead of the three-way handshake, whose handshaking packets would consume a significant bandwidth proportion in WSNs where the actual data packets may only be a few bytes in size. In addition, since WSN transfers often traverse a number of hops, TCP's end-to-end connection establishment before actual data transfer would also potentially add a lot of time to WSN network delays, a scenario that is undesirable in many mission-critical WSN applications.

3.2.5.1   Transport Protocols for WSNs

With the above challenges and requirements in mind, several transport protocols have been proposed for WSNs, a few of which are briefly described below. Details for each protocol can be found in the respective seminal paper for each protocol.

Congestion Detection and Avoidance (CODA) [32] CODA uses a threshold-based mechanism in which congestion is inferred if a given node's wireless channel load and buffer queue length exceed a certain preset value. When congestion is detected, the affected WSN node sends a back-pressure message to its neighbors, which in turn propagate the message, prompting neighboring nodes to reduce their sending rates. In the closed-loop, end-to-end mechanism, CODA nodes periodically probe the link before sending data, and detect congestion if the probes signal a link overload for a certain number of times. The nodes then set a bit in the header to notify the base station of overload, with the base station in turn prompting different sending nodes to decrease their rates. CODA has no in-built reliability mechanism and degrades sharply in performance when the number of nodes and total data rate increases [23].

Event to Sink Reliable Transport (ESRT) [1] In this algorithm, reliability in data transfer is in terms of events (set of packets associated with a given sensed event) rather than individual packets from each sensor. The algorithm uses a reliability metric, which it compares against the event reporting frequency of received packets per unit time pertaining to a given event, to determine the level of reliability of data transfer. After comparing the current reliability with the threshold, the ESRT algorithm reacts by alerting event measuring nodes to increase/decrease reporting frequency, depending on whether there is congestion or not. The challenge with ESRT is that the global setting of the frequency value (following broadcast-based notification from the base station) may not necessarily result in improved network performance since nodes in certain regions of the WSN may not always face the same congestion levels at different points in time.

TCP Implementation in Tiny TCP/IP [23] The implementation of standard suite of TCP/IP requires larger storage and cannot fit in memory of smaller devices such as sensors, set-top box, and medical devices; therefore, a smaller version of TCP/IP called tiny TCP/IP protocol suite is proposed [7]. The Tiny TCP/IP can be put in typical 512 kb flash memory or in in-built memory of microcontrollers. Tiny TCP/IP can be implemented in hardware as dedicated web server and has applications in internet of things (IoT) and WSNs including eMetering, lighting management, remote process control, interconnecting smart objects. This protocol improves on the conventional TCP/IP mechanisms in attempt to suite WSNs. The location of each sensor is assumed to be known, and the protocol thus uses an addressing scheme in which two octets identify the subnet of which node belongs, with the other two octets corresponding to the location within the subnet. Conventional TCP/IP comes with the burden of end-to-end retransmissions with the TCP protocol, and Tiny TCP/IP incorporates a form of TCP packet caching to

Fig. 3.28 Tiny TCP/IP:
Packets lost in transmission
are resent from intermediate
nodes that cached those
particular packets to reduce
retransmission overhead

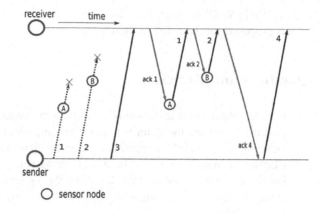

attempt to overcome this challenge. With TCP packet caching, intermediate nodes
cache packets between sender and receiver such that a lost packet may not have to
be fetched from the original sender, which could be many hopes away. Instead, it
could be recovered from an intermediate node's cache along the route. Figure 3.28
shows an example of how intermediate caches may be used to recover packets
between sender and receiver.

Intermediate nodes A and B are located between the sender and receiver and as
such store a cache of packets that are sent through them. Figure 3.28 shows the
packet recovery process in the event that the first two packets are lost during
transmission. Node A stores the first packet, and Node B stores the second packet.
In the event that both packets are lost, each is resent from its respective cache, i.e.,
packet 1 from node A, and packet 2 from node B. In the worst case, none of the
intermediate nodes has a copy of the missing packet and the sender has to resend
that packet.

Sensor TCP (STCP) [13] Most of STCP mechanisms are at the base station. The
protocol is based on three categories of packets: session initiation packets, data
packets and acknowledgement packets. Session initiation packets synchronize a
sensor node with the base station, and carry information such as number and type of
flows, transmission rates, and required reliability. Data packets in addition to car-
rying sensed data also maintain congestion information. The acknowledgment
(ACK) and negative acknowledgment (NACK) packets are used for feedback
mechanisms. Before a sender receives an ACK to signal data delivery, it maintains
a cache of the sent packet awaiting a possible retransmission process. Queue length
at the various nodes is used as a measure of the congestion levels, and the respective
nodes set a bit in the packet header to inform the base station of congestion.

A number of other WSN transport protocols include reliable multi-segment
transport (RMST), pump slowly, fetch quickly (PSFQ), GARUDA, Ad Hoc
Transport Protocol (ATP), Trickle and SenTCP among others. The major challenge
at this point is that many of the WSN protocols have been evaluated using simulations and are thus yet to see full evaluation in large-scale WSNs. This still leaves a

major void in WSN research field, as it is not easy to evaluate the usability of some of the proposed mechanisms in literature.

Questions and Exercises

1. Describe the typical architecture of a wireless sensor node and its main components. What are the main hardware building blocks?
2. Describe the principle of operation of an accelerometer.
3. Consider a mass sitting on a nonlinear spring (Fig. 3.29):
   Derive dynamic equation that describes the system above. Consider that the spring is nonlinear, meaning $F_{spring} = ky^2$. Linearize the model around equilibrium point y_0.
4. What is the difference between capacitive, piezoresistive, and piezoelectric accelerometer?
5. Describe the principle of operation of a pressure sensor.
6. What is the difference between piezoresistive and capacitance-based pressure sensors?
7. Describe how photodiode works. Draw p–n junction and describe the equilibrium state of electrons and holes across the junction.
8. Find a commercial photodiode on the web and get its important parameters from the datasheet such as gain, output current, terminal capacitance, cut-off frequency, etc.
9. What is a Hall effect and how magnetometers work based on the Hall effect?
10. Describe three types of chemical sensors.
11. Derive relationship between an induced voltage and differential surface stress on the cantilever beam. How the cantilever beam operates as a chemical sensor?
12. Use the following formula for a pressure sensor where capacitance changes based on the diaphragm deflection as $C = \int \int \frac{\varepsilon}{d - w(r)} r \, dr \, d\theta$. Assume that the

Fig. 3.29 Spring-mass system

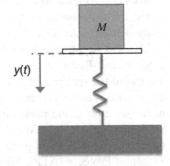

deflection follows a quadratic function pattern in terms of radial distance from the center of the circle r. Derive expression for capacitance C in terms of r for given dimension of a diaphragm with radius a.

13. What is the difference between general-purpose microprocessors and event-driven microprocessors? Describe design advantages in event-driven microprocessors in WSN applications.

14. What is the recommended size for the monopole antenna operating in 433, 915 MHz, and 2.4 GHz frequency range?

15. If the receiver antenna polarization is not perfectly aligned with the transmitter antenna polarization, calculate the power of the received signal including loss due to difference in polarization planes. Assume the angle between antennas' polarization is β, distance between sensor nodes is d, power of the transmitting signal is P, and the PLE is α.

16. Please describe the concept of sensor network asynchronous processor? What kind of operating system is required in this case?

17. Briefly explain how energy utilization may be minimized across the various layers of the WSN protocol stack.

18. TDMA, FDMA and CDMA are examples of fixed assignment protocols operating at layer 2 of the WSN protocol stack. Which of these protocols is best suited for a low duty cycle for power management? Briefly explain why.

19. Describe the mechanisms of operation of the following WSN transport protocols: CODA, ESRT, and Tiny TCP/IP.

20. In your own words, explain the difference between schedule-based and contention-based protocols for WSN media access control. What are the advantages and disadvantages of each in the context of a WSN?

References

1. O.B. Akan and I.F. Akyildiz, "Event-to-sink reliable transport in wireless sensor networks," *IEEE/ACM Transactions on Networking*, vol. 13, no. 5, pp. 1003–1016, Oct. 2005.
2. I.F. Akyildiz, W. Su, Y. Sankarasubramaniam and E. Cayirci, "Wireless sensor networks: a survey," *Computer Networks*, vol. 38, pp. 393–422, 2002.
3. E.H. Callaway, *Wireless Sensor Networks: Architectures and Protocols*, Auerbach Publications, New York, 2003.
4. M. Calleja, J. Tamayo, A. Johansson, P. Rasmussen, L. Lechuga, A. Boisen, "Polymeric cantilever arrays for biosensing applications," *Sensor Letters*, vol. 1, no. 1, pp. 1–5, 2003.
5. H.-M. Cheng, M.T. Ewe, R. Bashir, and G.T.-C. Chiu, "Modeling and control of piezoelectric cantilever beam micro-mirror and micro-laser arrays to reduce image banding in electro photographic processes", Journal of Micromechanics and Microengineering, vol. 11, pp. 487–498, 2001.
6. C.Y. Chong and S.P. Kumar, "Sensor Networks: Evolution, Opportunities, and Challenges," *Proc. of the IEEE*, vol. 91, no. 8, Aug. 2003, pp. 1247.
7. A. Dunkels, A.T. Voigt, J. Alonso, H. Ritter, and J. Schiller, "Connecting wireless sensor nets with TCP/IP networks," *Proc. Second International Conference on Wired/Wireless Internet Communications (WWIC2004)*, Frankfurt, Germany 2004.

8. V. Ekanayake, C. Kelly IV, and R. Monahar, "An ultra low-power processor for sensor networks," *ASPLOS'04*, October 2004, Boston, MA, USA.

9. P.M. Evjen and G.E. Jonsrud, "Short range devices antennas," *Texas Instruments Application Notes*, Available online: http://www.ti.com/lit/an/swra088/swra088.pdf (accessed September 2014).

10. J. Fraden, *Handbook of Modern Sensors: Physics, Designs, and Applications*, Springer-Verlag, New York, 2010.

11. M. Hempstead, M.J. Lyons, D. Brooks, and G.-Y. Wei, "Survey of hardware systems for wireless sensor networks," Journal of Low Power Electronics, vol. no. 1, pp. 1–10, 2008.

12. A. Hulanicki, S. Geab, and F. Ingman, "Chemical sensors definitions and classification," *Pure and Applied Chemistry*, vol. 63, no. 9, pp. 1247–1250, 1991.

13. Y.G. Iyer, S. Gandham and S. Venkatesan, "STCP: a generic transport layer protocol for wireless sensor networks," *Proc. 14th International Conference on Computer Communications and Networks*, 2005, pp. 449–454.

14. H. Karl and A. Willig, *Protocols and Architectures for Wireless Sensor Networks*, John Wiley and Sons, Ltd., 2005.

15. M. Kamarainen, M. Saukoski, M. Paavola, J.A.M. Jarvinen, M. Laiho, and K.A.I. Halonen, "A micropower front end for three-axis capacitive microaccelerometers," *IEEE Transactions on Instrumentation and Measurements*, vol. 58, no. 10, October 2009.

16. B. Koren, "Photodiodes – tutorial," *SPIE OEMagazine*, August 2001.

17. P. Lang, M.K. Baller, R. Berger, C. Gerber, J.K. Gimzewski, F.M. Battiston, P. Fornaro, J. P. Ramseyer, E. Meyer, H.J. Guntherodt, "An artificial nose based on a micromechanical cantilever array," *Anal. Chim. Acta*, 1999, 393, pp. 59–65.

18. J.E. Lenz, "A review of magnetic sensors," *Proceedings of the IEEE*, vol. 78, no. 6, June 1990.

19. G. Mao, B.D.O. Anderson and B. Fidan, "Path loss exponent estimation for wireless sensor network localization," *Computer Networks*, vol. 51, no. 10, pp. 2467–2483, July 2007.

20. Y. Nemirovsky, A. Nemirovsky, P. Muralt, and N. Setter, "Design of Novel Thin-Film Piezoelectric Accelerometer," *Sensors and Actuators A: Physical*, vol. 56, no 3, September 1996.

21. F. Primdahl, "The fluxgate magnetometer," *Journal of Physics E: Scientific Instruments*, vol. 12, no. 4, 1979.

22. R. Puers and S. Reyntjens, "Design and processing experiments of a new miniaturized capacitive triaxial accelerometer," *Sensors and Actuators A: Physical*, vol. 68, no. 1–3, pp. 324-328, Jun 1998.

23. A. Rahman, A. El Saddik, and W. Gueaieb, "Wireless Sensor Network Transport Layer: State of the Art," *Sensors, Lecture Notes in Electrical Engineering*, Springer-Verlag, vol. 21, 2008, pp. 221–245.

24. T. Seiyama, A. Kato, K. Fujiishi, M. Nagatani, "A new detector for gaseous components using semiconductive thin films," *Analytical Chemistry*, vol. 34, no. 11, pp. 1502–1503, 1962.

25. S. Senturia, *Microsystems Design*, Kluwer Academic Publishers, Norwell, MA, 2001.

26. M. Sheets, F. Burghardt, T. Karalar, J. Ammer, Y.H. Chee, and J. Rabaey, "A power-managed protocol processor for wireless sensor networks," *2006 Symposium on VLSI*, June 2006.

27. K. Sohraby, D. Minoli, and T. Znati, *Wireless Sensor Networks: Technology, Protocols and Applications*, John Wiley and Sons, 2007.

28. J.R. Stetter, W.R. Penrose, and S. Yao, "Sensors, chemical sensors, electrochemical sensors, and ECS," *Journal of The Electrochemical Society*, vol. 150, no. 2, pp. S11-S16, 2003.

29. TinyOS Community Forum at http://www.tinyos.net/.

30. S.C. Thompson, "Tutorial on microphone technologies for directional hearing aids," *The Hearing Journal*, vol. 56, no. 11, November 2003.

31. M. Tomic, P.T. Sullivan, and V.K. Mcdonald, "Wireless, acoustically linked, undersea, magnetometer sensor network," *OCEANS 2009, MTS/IEEE Biloxi - Marine Technology for Our Future: Global and Local Challenges*, Oct. 2009.

32. C.-Y. Wan, S. B. Eisenman, and A. T. Campbell, "CODA: Congestion Detection and Avoidance in Sensor Networks," *Proc. the 1st International Conference on Embedded Networked Sensor Systems*, 2003, pp. 266–279.

33. B.A. Warneke and K.S.J. Pister, "An ultra-low energy microcontroller for smart dust wireless sensor networks," *Proc. 2004 IEEE International Solid-State Circuits Conference*, San Francisco, CA, 2004.

34. G. Wu, H. Ji, K. Hansen, T. Thundat, R. Datar, R. Cote, M.F. Hagan, A.K. Chakraborty, and A. Majumdar, "Origin of nanomechanical cantilever motion generated from biomolecular interactions," *Proc. National Academy of Science*, USA, vol. 4, 2001, pp. 1560–1564.

35. G. Yu, K. Pakbaz, and A.J. Heeger, "Semiconducting polymer diodes: Large size, low cost photodetectors with excellent visible-ultraviolet sensitivity," *Applied Physics Letters*, vol. 64, no. 25, pp. 3422–3424, Jun 1994.

36. F. Zee and J. Judy, "MEMS chemical gas sensor using a polymer-based array," *Proc. Transducers 1999 - The 10th International Conference on Solid-State Sensors and Actuators*, June 1999, Japan.

37. B. Zhai, S. Pant, L. Nazhandali, S. Hanson, J. Olson, A. Reeves, M. Minuth, R. Helfand, T. Austin, D. Sylvester, and D. Blaauw, "Energy-efficient subthreshold processor design," *IEEE Transactions on Very Large Scale Integration (VLSI) Systems*, vol. 17, no. 8, Aug. 2009.

38. F. Zhao and L. Guibas, *Wireless Sensor Networks*, Elsevier, 2004.

39. W. Zhou, A. Khaliq, Y. Tang, H.-F. Ji, R.R. Selmic, "Simulation and design of piezoelectric microcantilever chemical sensors," *Sensors and Actuators A*, vol. 125, no. 1, pp. 69–75, October 2005.

40. www.atmel.com

41. www.ti.com

42. *SensEdu* – an e-learning material, http://www.sensedu.com/

43. http://www.freescale.com/

44. http://www.ams.com

45. http://www.openmusiclabs.com

Chapter 4
Security in WSNs

Security in wireless sensor networks (WSNs) is centered on at least the following six fundamental requirements, namely: authentication, confidentiality, integrity, reliability, availability and data freshness [4, 39, 52, 62]. In this chapter, we describe these requirements, the different kinds of attacks that aim to compromise these requirements (and hence the security of a WSN) and the defense mechanisms that can be employed to overcome these attacks. While many security concepts hold true for different kinds of computer networks, WSNs have certain properties that make them particularly susceptible to attack relative to the traditional computer networks (e.g., wired networks). These properties do not only make possible a range of attacks not seen in regular computer networks, but also make a number of conventional defenses unsuitable for WSNs. We first describe examples of these unique properties and how they give rise to WSN-specific security challenges before delving into the WSN security requirements, attacks and defenses. The last part of the chapter covers WSN reliability including fault awareness and fault detection in WSNs.

4.1 Why WSNs are Predisposed to Attacks?

In this section, we describe unique properties that predispose WSNs to various types of attacks.

Resource Limitations Many security mechanisms employed in computer networks rely on some form of cryptography. While public key cryptography is in many cases a more versatile security scheme relative to private key cryptography [52], the limited memory and processing power of the WSN nodes mean that the former method can only be used subject to very careful optimization of the algorithms at both the design and implementation levels [64]. This challenge does not arise with traditional computer networks.

Large-scale Deployment Sensor networks typically contain very many nodes (e.g., thousands of nodes) that work in collaboration to achieve the purpose of the network. This means that the two-party functionality seen with many security protocols may not always be suited for WSNs [52]. Further, the large number of nodes sharing sensitive information requires that security designs have to be cognizant of the fact that a single compromised node could leak sensitive information about the entire network.

Open Deployment Because sensors are typically deployed in outdoor unattended environments, an attacker could very easily have physical access to them and extract sensitive information (e.g., security keys) from the motes. This is in contrast to traditional computer networks where it is for the most part safe to assume threat models in which an adversary will have very low likelihood of physically accessing a node.

Wireless Connectivity Because WSNs communicate via wireless communication channels, the adversary only needs to tune into the frequencies being used for communication to be able to eavesdrop on traffic being exchanged between the nodes. Different from a wired computer network, therefore, security mechanisms for WSNs need to take these kinds of attacks into consideration.

The unique attributes such as those described above make possible a number of attacks against WSNs that call for defenses, which are for the most part specifically tailored to WSNs.

4.2 Security Requirements

The fundamental security requirements of a WSN are given below.

Authentication Authentication enables a node to confirm that the identity of another node with which it communicates is as claimed. This helps a node to verify the origin of packets sent to it, ruling out the chance that spoofed packets or packets maliciously injected into the wireless communication channel may be mistaken for genuine packets. A message authentication code (MAC) attached to a message can be used to verify the origin of a message [64].

Confidentiality Confidentiality stipulates that information is only accessed by nodes which are supposed to access it. Confidentiality is ensured by encrypting the packets being sent out at the originating node so that they are decrypted at the receiving node. Depending on the application, encryption could be applied to the data part of a packet or the full packet (including the header) [64]. Encryption of the full packet helps obfuscate the node identities (which are located in the header) which helps minimize the chances that a given node's identity could be spoofed by an eavesdropper.

Integrity Integrity ensures that a message sent between two nodes is not modified by an adversary. If for instance routing or clock synchronization packets exchanged between nodes are modified by a malicious entity, the entire network could come to a halt. Data aggregation (i.e., the combination of readings obtained by multiple sensors) is another WSN service that heavily relies on the assumption that data being exchanged between WSN nodes is not modified maliciously [52]. A keyed checksum (e.g., a MAC) can be used to determine whether messages have been modified [64]. Figure 4.1 illustrates how a MAC appended to a message at the sending node can be used by the receiver to determine the integrity of the message. Taking the message and a key as input, the sender uses a MAC algorithm to compute the MAC, which is then appended to the message before it is sent out to the receiving node. At the receiving node, the message and the key (same as the key which was used at the sending node) are input to the MAC algorithm so as to compute a MAC. If this MAC matches with the one retrieved from the message, then the message is confirmed not to have been tempered with.

Availability Availability indicates that the WSN must be functional at all times and provide services whenever needed. One way in which availability can be achieved is by ensuring reliability at both the node and network level, which can in turn be achieved by building fault tolerance into the individual nodes and the network [27, 29, 57]. Designing the system against different kinds of denial of service attacks is also crucial for the guaranteed availability of the WSN.

Data Freshness WSN systems typically either continuously sense and forward data from the environment, or forward data in response to a certain event. In both cases, it is crucial that data sensed by the nodes reaches the base station as soon as possible (i.e., when is still fresh). This does not only minimize the likelihood that replayed packets sent by an attacker could be mistaken for legitimate packets, but also ensures that the right kinds of interventions can be undertaken to react to the events sensed by the network. In a target tracking application for instance, the target can only be tracked if current data is forwarded to the base station. Data freshness is in general achieved through reliable transport and routing schemes that minimizes packet losses and delays.

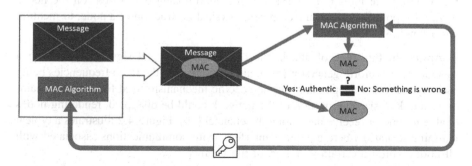

Fig. 4.1 Using a MAC to verify message integrity

4.3 WSN Attacks and Defenses

There are several approaches to the categorization of attacks on WSNs. One of these approaches categorizes attacks based on whether they disrupt the functionality of the network or not. Attacks which disrupt network functionality are called active attacks (e.g., network jamming attacks [59]), while those which do not disrupt network functionality are referred to as passive attacks (e.g., packet eavesdropping attacks [61]). Another common way to categorize WSN attacks is to classify them based on whether they are internal (i.e., launched by nodes which are part of a WSN) or external (i.e., launched by nodes or devices that are not part of the network). The Sybil attack [36] is an example of an internal attack, while a network jamming attack [59] is an example of an attack which would typically be launched by an external agent (i.e., an external attack). It is noteworthy that an attack launched by an external entity which gets authorization to access the network and then exploits the privileges to launch attacks would be classified as an internal attack [64].

The most prominently used approach for the categorization of WSN attacks in the literature distinguishes between attacks based on the layer of the communication architecture which the attack targets. We adopt this approach in this work, since it gives a fine-grained view of the aspects of the different algorithms and (or) protocols that each attack seeks to leverage. For each layer of the network model, we first discuss the potential attacks before delving into the different defenses that could be used to thwart the attacks.

4.3.1 Physical Layer Attacks

We discuss here several physical layer attacks, including tempering, jamming, and eavesdropping and traffic analysis.

Tempering Given physical access to a WSN node, the attacker could temper with the node in several ways, which include reprogramming the node with the aid of easily accessible tools (e.g., UISP [23]), compromising data stored on the nodes (e.g., encryption keys) and the complete physical destruction of a node, to mention but a few [23].

Jamming In this type of attack, the adversary sends out signals (e.g., using a specialized waveform generator [59]) that interfere with the radio frequencies being used by the WSN. Depending on the specific mechanism used to launch the jamming attack, a sizeable portion of the network could be disrupted, rendering nodes unable to send or receive data along the channel [59]. Figure 4.2 illustrates a typical jamming scenario where a jammer interferes with communications associated with all nodes within a certain radius, r, of the jammer.

Fig. 4.2 A jamming attack
disrupting all communications
between nodes within a radius
r of the jamming node

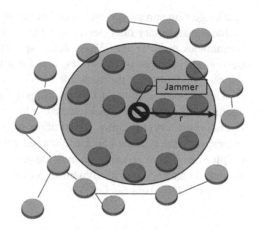

Eavesdropping and Traffic Analysis Since WSN signals are broadcast in the air, an adversary within the range of the signals could listen to the transactions going on in the wireless channel with the aid of antennas that cost as little as $20 [10]. From data captured during eavesdropping, the adversary could directly extract the message content, or carry out traffic analysis to make inferences such as the location of the base station, which could in turn inform more targeted attacks.

4.3.2   Physical Layer Defenses

One potential defense against tempering is the design of temper-proof WSN nodes (e.g., by encasing them in a physically sturdy package [62]). The challenge with this option, however, is that the cost of each mote could tremendously increase, making the deployment of large WSNs very expensive. Under the assumption that an adversary accessing a WSN node will have to move it (e.g., from one place to another), building location awareness into the applications running on the motes could be used to defend against some physical attacks. If a node is determined to have been moved (e.g., based on accelerometer or GPS sensor readings), its data could for instance be flagged by other nodes as being potentially compromised. Alternatively, the node could be configured to delete sensitive information from memory if movement is detected [62].

Jamming and eavesdropping attacks could be mitigated through the use of spread spectrum communication. The philosophy behind spread spectrum communication is to distribute the communication channel over a large number of frequencies, making it very expensive for the adversary to jam or eavesdrop on all the frequencies. In one spread spectrum technique (called frequency-hopping [66]), senders rapidly switch between frequencies using a pattern known to the receivers. This way, an adversary would not know which frequency to jam or eavesdrop while the receiver would be aware of the variations in transmission frequencies. The challenge with spread spectrum communication in broadcast systems such as WSNs is that the adversary only needs to compromise one node to determine the range of frequencies used by the WSN. Also, spread spectrum functionality increases the power requirements of the nodes, and for a large WSN, can dramatically increase the cost of deploying the network.

The ultimate solution to eavesdropping/traffic analysis on the communication channel and tempering attacks aimed to infer information stored on a captured node is encryption/cryptography. Encryption may be applied to the data stored on the motes, or to the packets being exchanged between nodes over the channel. Because cryptography is a solution to a wide range of other attacks, we will discuss it separately and address its dynamics and the different factors affecting its applicability in WSNs (see Sect. 4.4).

4.3.3 Link Layer Attacks

MAC Protocol Violations MAC protocols generally help ensure that the sensors in the network efficiently use the shared communication channel. When a given node violates the MAC protocol mechanisms (e.g., by sending data during a time slot when another node is supposed to be sending), packet collisions occur. Depending on the extent of the violation, the collisions could result into a wide range of issues, including corruption of the data in the packets, unfair bandwidth usage, and in the worst case, total denial of service if the malicious sender continuously occupies the channel and (or) the attacked nodes continually attempt to retransmit corrupted packets [62].

MAC Identity Spoofing When a node broadcasts data on a WSN, its MAC identity can be accessed by all nodes sharing the communication channel. Given MAC addresses of nodes on a WSN, a malicious node (that may be within range of the network without necessarily being part of the WSN) could use these identities to masquerade as any of these nodes. If, for example, the malicious node masquerades as an aggregation point, it could leverage this role to access privileged resources of the WSN [61]. The Sybil attack [36]—an attack in which a malicious node on the WSN presents different identities to the network at different points in time or cycles through multiple identities to create the impression that they are all simultaneously present on the network—is realized in WSNs through MAC identity spoofing.

4.3.4 Link Layer Defenses

Error correcting codes [51] can be employed to address data corruption issues caused by packet collisions if the extent of collision is low to moderate (e.g., if collisions are comparable to those seen due to probabilistic errors [62]). However, this approach would not handle a heavy volume of collisions as it would require a considerable amount of resources [62]. Incorporating rate-limiting in the MAC protocol to ignore requests beyond a certain threshold of requests and the use of time-division multiplexing to limit the amount of time that nodes can use the channel are the other potential defenses against attacks which overload the channel through violations of the MAC protocol [62].

MAC identity spoofing can be tackled by associating each identity with a secret key, making it impossible for the adversary to use a given address without having its associated key [62]. Under the assumption of an immobile WSN, position information (registered when the network is set up) can be used in conjunction with MAC identities to confirm that a node is not using a spoofed address [36]. Other techniques, such as node registration (at a central server which can be polled to verify the identities of nodes in the network), and code attestation (e.g., to remotely determine the legitimacy of node by comparing its memory contents to those of known legitimate nodes) have also been proposed to address these kinds of attacks [36, 50].

4.3.5   Network Layer Attacks

Routing Manipulation Attacks The routing tables used by nodes to forward packets in a WSN can be poisoned in a number of ways. For example, a malicious node may modify or spoof route update packets before forwarding them to the other nodes in the network. As a result, the nodes receiving these updates may direct traffic along routes determined by the attacker, which could in turn result into congestion and collapse of the network [61]. Some specific types of route-poisoning attacks include:

(a) *Sinkhole* attack: In this attack, the adversary manipulates routing information to lure a large number of nodes into routing their traffic via a node controlled by the adversary [26, 62] (we also alternately refer to this node as the malicious or attacking node).

(b) *Wormhole* attack: This attack is centered on route manipulations designed to make two distant malicious nodes appear to the other nodes to be much closer to each other than is actually the case. Figure 4.3 illustrates the concept of a wormhole attack. The compromised nodes 11 and 2 have a direct (wormhole) path which is much shorter (in terms of number of hops) than other potential paths between the two parts of the network. Nodes 12 and 1 are very likely to communicate via the wormhole path since this path creates the illusion that these notes are very close to each other (in fact, much closer to each other than they actually are). Without the wormhole attack, these two nodes would ordinarily have communicated via the much longer route via nodes 11, 8, 6, 7, 5, 4 and 2.

Fig. 4.3 A wormhole attack
in a WSN

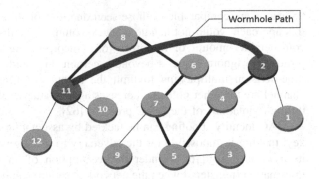

If the adversary strategically locates the wormhole in such a way to fool the
other nodes that they are just a few hops away from the base station via the
wormhole, another malicious node placed between the wormhole and the base
station can then manifest as a sinkhole [26]. Besides the sheer destabilization
of the routing process, attacks such as the sinkhole and wormhole attacks can
also be used as a vehicle for eavesdropping given the high amount of traffic
traversing the nodes being controlled by the adversary.

(c) *HELLO Flood* attack: This attack basically involves the attacker sending
HELLO packets to various nodes in the network to dupe them into classifying
the attacking node as their neighbor [26, 62]. Because some of the nodes may,
in fact, be very far away from the malicious node, the adversary may have to
use a high-powered transmitter (e.g., laptop class device) to achieve the range
of transmission required to deliver the HELLO packets. If this malicious node
(now perceived to be a neighboring node) broadcasts a low cost route to the
base station, WSN nodes may attempt to send data via this route, potentially
resulting into failed data transfers, retransmissions and channel congestion
(since the offending node is actually not in radio range with many of the nodes
attempting to send data via it.). Attacks such as the sinkhole and wormhole
attacks could be realized with the aid of a HELLO Flood attack if the malicious node(s) is (are) located in such a way to enable successful data transfers
from the victim nodes to the malicious node(s).

(d) *Acknowledgement Spoofing*: When a node *A* sends out packets to node *B*,
routing algorithms used by WSNs require that *B* (explicitly or implicitly)
sends some form of acknowledgement to *A* if data from *A* is indeed received
by *B* [26]. A malicious node *C* that is aware of data being sent from *A* towards
B (awareness of the transaction arises out of the broadcast nature of the
channel) can spoof the identity of node *C* and send acknowledgement information towards *A*, which would then believe that the acknowledgement indeed
originated from *B*. This kind of attack could, for instance, fool node *A* to
believe that node *B* is alive yet it is in actual sense dead. With node *A* having a
wrong view of the network topology, an unstable routing process could result.

(e) *Selective Forwarding*: Rather than forward all received messages as is sup-
    posed to be the case in multi-hop networks, a malicious node launching this
    attack may only forward a subset of the messages [26, 61]. This kind of attack
    could, for instance, be launched by an adversary who is interested in sup-
    pressing the propagation of traffic originating from a certain node or subset of
    nodes. In the extreme case, a malicious node launching this attack could drop
    all packets received by it, creating what is termed as a black hole [26].
    Figure 4.4 illustrates a black hole in the middle of a WSN. The dark colored
    nodes route their traffic via the black hole, which drops all packets and denies
    them access to the base station.
    A notable aspect of the black hole variant of selective forwarding attacks
    however is that neighboring nodes may assume the malicious node to have
    failed, prompting them to use alternative routes that do not go through the
    malicious node. More subtle forms of this attack may hence be more attractive
    for an adversary who is keen to avoid triggering such route modifications that
    may result into the exclusion of the malicious node. It is noteworthy that
    selective forwarding attacks require that the malicious node is positioned in
    such a way to be traversed by the traffic to (or from) the node that the attacker
    seeks to target. To meet this requirement, these types of attacks may leverage
    aspects of other attacks, such as the sinkhole, wormhole and Sybil attacks to
    mention but a few.

Routing Table Overflows Route-poisoning can also be induced by overflowing
the routing tables in the victim nodes. In particular, by continually sending void
routing information to the WSN, an adversary could ensure that nodes in the
network have bogus routing information in their routing tables, with little or no
room available in the node buffers for correct routing information [61].

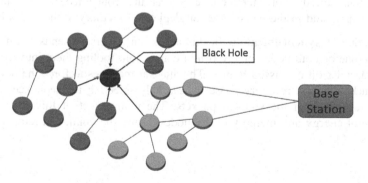

Fig. 4.4 Black hole illustration: all messages received at the black hole are not forwarded to their
destinations

4.3.6 Network Layer Defenses

The use of multiple paths for data transfer can limit the effects of some route manipulation attacks (e.g., selective forwarding, and the sinkhole and black hole attacks), since the damage to the system would be restricted to only the traffic traversing a subset of the paths. The challenge with this approach, however, is that the maintenance of potentially redundant routes may have its toll on network resources, let alone the scheme not being feasible in sparse networks which have very few possible routes [46]. Authentication schemes that only update routing table entries after the node originating an update is verified are another defense against these kinds of attacks [61]. At the message level, a message authentication code (MAC) attached to each message (including routing-specific messages) can be used to determine if routing information could have been altered [62]. Against wormhole attacks, precise clock synchronization between communicating nodes could be leveraged to estimate the distance traveled by each packet and rule out the possibility that a given malicious node is much further than it claims to be [61]. Geographic routing protocols that use knowledge of each node's location information as additional information to determine the source of a given packets can help overcome attacks such as the HELLO floods or acknowledgement spoofing [46].

4.3.7 Transport Layer Attacks

Denial of Service As part of the mechanisms to ensure end-to-end reliability, transport layer protocols in WSNs (see survey in [11]) maintain state information (e.g., information on the status and identity of each active connection). When an adversary opens up a large number of connections in a short time, the amount of information stored about the connections at the concerned node(s) increases tremendously, and in the worst case can deplete all memory at the nodes [62].

Connection Desynchronization In this attack, an adversary sends spoofed messages to one or both nodes involved in a connection, fooling them into requesting for retransmission of missed frames. The attacker manages to force the nodes into this synchronization recovery phase by carefully selecting the sequence numbers and control flags of the spoofed packets. The main effect of this attack is the wastage of energy and memory of the nodes during the continued retransmission requests [46].

4.3.8 Transport Layer Defenses

A possible defense against *connection desynchronization* attacks is the authentication of all packets being exchanged during a connection. This helps ensure that packets from a malicious sender cannot be confused with those being sent between the genuine nodes. For the DoS attacks against the transport layer, one proposed solution is to force each node to commit to every connection it initiates, e.g., by having it first solves a puzzle [62]. The puzzles would have the undesired impact of consuming a node's resources, but would make it computationally challenging for a node to initiate a very large number of connections and leave them unclosed [62].

4.3.9 Application Layer Attacks

Attacks on Data Aggregation Process Data aggregation, a service provided by the application layer, ensures meaningful combination of data from the sensors (e.g., by eliminating erroneous readings) to enable the accurate estimation of the sensed environment [52]. Possible attacks on this service include the malicious modification of data before it is forwarded to the base station and the complete disruption of the service by, for instance, strategically placing a black hole in the WSN. With the base station having wrong information about the sensed environment, the network could then be compromised in several other ways if the base station triggers actions based on the wrong information fed to it [61].

Clock Desynchronization Clock synchronization, another service provided by the application layer, typically has to be carried out with very high accuracy [14, 52]. Attacks on this service (i.e., clock desynchronization attacks) can be launched by sending malicious beacon packets into the network, forcing some nodes to adjust their clocks and get out of sync with the other nodes [61]. For example, in WSN applications in which time series data is used to make velocity estimates (e.g., see [7]), clock synchronization problems would result in wrong velocity computations, defeating the very purpose of the network. The identification of duplicate readings from different nodes detecting the same event is another WSN operation which heavily relies on the sensor clocks being closely synchronized [14, 25].

Selective Forwarding This attack is similar to the one already described under the network layer, except that the attacker or compromised node makes forwarding decisions based on the contents of the packets as opposed to the addresses of the nodes involved [61].

4.3.10 Application Layer Defenses

Attacks on data aggregation and clock synchronization are essentially attacks on the integrity of data being exchanged between nodes, meaning they have to be addressed through authentication as data finds its way through the network. The selecting forwarding attack is addressed through encryption schemes that obfuscate the data [61].

4.4 Cryptography in Sensor Networks

Cryptography is at the heart of many WSN security mechanisms designed to ensure data confidentiality and authenticity. The two categories of cryptography—symmetric (or private) and asymmetric (or public) key cryptography—are both used in several cryptography schemes proposed for WSNs. For each of symmetric and asymmetric key cryptography, this section gives an overview of the design considerations and operational dynamics of the different schemes that have been evaluated or proposed for WSNs. A more detailed treatment of these topics can be found in [62, 64].

4.4.1 Symmetric Key Cryptography in WSNs

In a symmetric key system, the sender encrypts a message using a secret key that is also known to the receiver of the message. On receiving the cipher text (i.e., encrypted message), the receiver uses the same key to decrypt the message. A major requirement of symmetric key systems is that the keys remain secret, and be changed frequently to minimize the possibility of success of attacks. Further, for large networks, it is crucial that the distribution of keys between the sensor nodes be done as efficiently as possible because the number of key-pairs can be very large. To meet the above two requirements (i.e., key security and efficient key distribution), there is a wide range of techniques (i.e., key management schemes) whose applicability to WSNs has been widely studied in the literature. We briefly discuss these schemes in the rest of the section.

The simplest approach to key establishment in WSNs would be to use a single key that is known to all nodes on the WSN. If one node has to communicate with another, it only has to encrypt data using the global key, so that the receiver can decrypt the data using the same key. The challenge with this approach is that a single compromised node results into the entire WSN being compromised.

Centralized key distribution (e.g., the scheme used in [20]) presents a more secure alternative to the above scheme as each node only shares its key with the base station. Communication between any two WSN nodes begins with each of them communicating with the base station, which sends the symmetric key required for the two nodes to communicate. A weakness with this scheme is that it could incur a huge communication overhead, e.g., when two neighboring nodes have to first communicate with a distant base station before exchanging data with each other. This scheme also faces a single point of failure vulnerability since a breach of the base station exposes the keys for each pair of nodes on the WSN [64].

Fig. 4.5 Pairwise key used for communications between each pair of nodes on the network

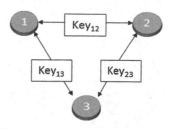

A naïve approach of solving the single point of failure problems seen with the above two approaches is to pre-distribute the keys, i.e., preload each node pair with a private key before deploying the nodes on a WSN (Fig. 4.5). Assume a WSN having N nodes. Since each node will have to store $N - 1$ keys, a total of $0.5N(N - 1)$ keys will have to be stored in the WSN. As N increases, this number will grow very fast, potentially imposing significant memory constraints on the WSN [64]. To reap the benefits of key pre-distribution while minimizing resource consumption, there is a wide range of key distribution schemes that have been proposed in the literature [3]. We briefly discuss these next.

Random Key Pre-Distribution (RKP) [15] In this approach, each WSN node is preloaded with a key ring, which is a set of keys randomly selected from a large pool of keys. With a certain probability, two nodes will share a key, and hence be able to communicate directly using that key. Figure 4.6 illustrates how RKP works. The nodes numbered 1, 2, 3, 4 and 5 randomly choose keys from a large pool of keys. Node 1's key ring has the keys a, b and c while nodes 2, 3, 4 and 5, respectively, have key rings comprised of the keys p, q and y; a, y and q; p, x and y; and x, y and p. Nodes numbered 1 and 3 have the key a in common, and can hence use it for their pairwise communications. Similarly, nodes 3 and 4 can use the key y for their communications. The other nodes use their common keys for communication in the same way.

It is noteworthy that the absence of direct connectivity between every pair of nodes will not prevent communication between any two nodes because data transfer between a pair of nodes only requires that a path exists between the two nodes via other nodes which share keys pairwise. For example, node 3 can communicate with node 2 through node 5 which shares a key with both nodes. Similarly, node 1 can communicate with node 4 via node 3.

With a WSN of 10,000 nodes, it was shown in [15] that key rings of just 250 keys can guarantee "almost-certain" connectivity for the entire network. This translates into a total of approximately 2.5 million keys, which is a significant reduction in memory requirements if compared to the approximately 50 million keys required (i.e., $0.5N(N - 1)$) for a similarly sized network using the previously described scheme.

Variants of RKP Because the basic RKP scheme requires that two nodes only to have a single shared key to set up a secure communication link, the compromise of a single key ring (i.e., capture or compromise of a single node) exposes a large

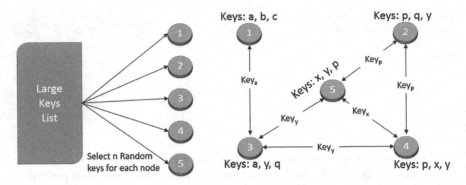

Fig. 4.6 Illustration of random key pre-distribution

number of nodes to attack. The q-RKP scheme [8] seeks to improve the resilience of a WSN to attack by requiring that two nodes should share q keys, where $q > 1$ to be able to set up a communication link. As q increases, the chance that an attacker who compromises a set of keys can break a given link decreases exponentially [8]. An augmentation of the q-RKP scheme uses multipath reinforcement to improve security [8]. In particular, after the initial secure links have been set up, this scheme requires that the nodes update the keys with random values using multiple independent links. The use of multiple links helps limit the impact of node capture attacks on the network. The use of multiple independent links helps ensure that an attacker listening to the network would have to listen to multiple links to be able to extract enough information to compromise the network. In a location-dependent version of RKP [1, 23], key rings are constituted based on the locations of the sensors in such a way that neighboring nodes have more keys in common than distant nodes. The advantage of this is that the effects of node compromise are constrained to a local area. Several other derivatives of RKP have been proposed (e.g., see [12, 24, 32, 41]) with the major motivation being the improvement of the security, memory requirements and network connectivity properties of the base key pre-distribution protocol.

Deterministic Key Distribution A shortcoming of the RKP protocols is that there is a possibility of certain nodes being isolated from their neighbors due to a lack of shared keys. This weakness follows from the non-guaranteed connectivity depicted by the underlying random graph model on which the base RKP protocol is based. Deterministic key pre-distribution schemes overcome this challenge by imposing some requirements on the extent of connectivity between nodes. In one deterministic key pre-distribution scheme (see [31]), a strongly regular graph is used as the underlying model, guaranteeing that any two adjacent nodes have connectivity to a certain number of neighbors and every two nonadjacent nodes have connectivity to a certain number of neighbors. In another approach to deterministic key pre-distribution (e.g., see [63]), the WSN is modeled using a multidimensional grid in which nodes along the same dimension are able to establish shared keys while nodes with dimension mismatches can negotiate keys through indirect connections.

4.4.2 Asymmetric Key Cryptography in WSNs

To ensure message confidentiality through asymmetric key cryptography, a sender uses the receiver's public key to encrypt a message while the receiver uses their private key to decrypt the message. The private key is only known to the receiver, while the public key can be accessed by any entity which wishes to send a message to the receiver. Asymmetric cryptographic schemes are, however, resource-intensive, as they involve complex computations such as the modular exponentiation of large numbers [9]. In a WSN setting, while the decryption process can for some applications be assumed to be done at the base station (which can be assumed to have a significant amount of resources) [20], the encryption step has to be done on the WSN nodes, which are inherently resource-limited. Asymmetric cryptography, however, has a number of advantages that make it appealing for WSN applications. One major advantage of asymmetric key cryptography relative to symmetric cryptography in WSNs is that it is robust against node capture attacks since a captured node would only reveal its private key and no other information. Other advantages of these schemes include their scalability (since they perform the same way regardless of the number of nodes in the network [9]), and their ease of revocation of compromised key-pairs [9].

Early efforts towards asymmetric cryptography in WSNs worked around the resource-intensiveness of the assymetric algorithms via a scheme which simulates an asymmetric system through a delayed disclosure of symmetric keys in which keys are revealed sequentially over time (scheme is presented in [40]). As argued in [20], this delayed disclosure of keys may require nodes to store multiple messages before they can be authenticated, which can, in turn, take its toll on the resource-constrained nodes. Another challenge posed by the emulated asymmetric cryptography scheme in [40] is that it requires frequent transmission of the keys and key management messages, which has implications on both the energy requirements and security of the network.

On basis of these factors, Gaubatz et al. [20] argue that emulated asymmetric cryptographic schemes in WSNs eliminate many of the benefits of symmetric key cryptography and propose the use of customized hardware to support the expensive asymmetric key operations. In particular, they show using customized hardware that two well–known asymmetric cryptographic schemes—the NtruEncrypt and Rabin algorithms—are feasible for WSN systems. Software-based approaches that minimize the computational load on the nodes (e.g., by among other factors selecting a low value of the exponentiation base, e, used in the cryptographic computations) have also been found to enable the implementation of public key cryptography in WSNs. For example, in [58], an RSA-based method that uses $e = 3$ is shown to work well for the UC Berkeley MICA2 motes. A much wider range of cryptographic schemes is studied in [42] with the authors finding that with careful design, asymmetric cryptography is much more feasible for WSNs than earlier thought.

4.5   Faults in WSNs

Reliable performance of WSNs is necessary in safety-critical applications, such as process control, chemical agents monitoring, radiation contamination monitoring and detection, medical applications and others. Here, we present an overview of statistical data analysis and signal processing techniques that are used to detect sensor faults. Sensor network systems have four levels of abstraction as listed below, and failures can occur at each level.

4.5.1   Fault-Aware WSNs

Fault-aware WSNs have ability to detect, isolate, and mitigate faults at each system level including sensors (components), sensor nodes, and network (system) level. At each level at which faults can occur, fault-aware WSNs are required to detect failures, isolate them and mitigate their effects. Such design improves robustness and reduces the need for hardware redundancy. However, the tradeoffs are usually in higher computation and communication costs that are incurred during fault analysis and mitigation.

Component Level Each sensor node has multiple components, such as sensors, actuators, a power supply, a radio and a microcontroller. Sensors and actuators directly interact with the physical environment and are subjected to noise, aging and errors due to manufacturing defects. The depletion of batteries can cause the sensor nodes to crash. A node with failed radio is equivalent to a crashed node even if the other sensing and power subsystems operate well. Various methods have been proposed to mitigate sensor failures including novel fault tolerance schemes [28, 34] and detection of faulty sensors using Bayesian algorithms [13, 30].

Node Level At this level, data processing includes converting the analog sensor data into digital data. Data encoding and decoding techniques and data compression can be implemented to increase data rate. A node failure may occur because of the physical damage caused by random airdrop or exposure to harmful chemicals. Crash faults in sensor networks are identified in [10]. Fault identification algorithms can be implemented on individual sensor nodes. This increases the storage and computational cost at the node level. On the other hand, this method has the advantage of saving the communication cost required to transmit all the sensor data to the base station. Radio is the most power-consuming component of the sensor node. Applications that perform data aggregation or data processing from multiple

Table 4.1 Example of sensor node energy consumption for Telos-B node	Operation	Energy consumption
	Transmit one bit (radio)	2950 nJ/m
	Receive one bit (radio)	2600 nJ/m
	Execute one operation (processor)	4 nJ/operation

nodes require the base station to inform them about the failed sensor nodes. Such applications would need to remove the failed nodes readings from collaborative operations of sensor nodes. Therefore, there is a two-way communication cost involved in centralized fault detection algorithms. Table 4.1 lists the energy consumed by the Telos-B mote radio and processor [53].

Network Level The nodes form a bidirectional communication link with their peers to achieve a network of interacting sensor nodes. The network layer of a WSN is responsible for efficient communication in the transmission medium. Inability to perform efficient routing tasks may constitute to congestion in the network. Lightweight and energy-efficient transport protocols that can support any generic application have been developed in [49, 55].

System Level For proper functioning of any system, accurate coordination of the distributed actions performed by the diverse system components is needed. The system level analysis involves a centralized method of implementation in which data from all the nodes are gathered at the base station prior to analysis. Algorithms are implemented at a local computer or the data might be transported to remote users via wide area network for analysis and processing. The advantage of this implementation method is that local or remote computers are equipped with more computational and storage resources than a tiny sensor node.

4.5.2 Sensor Faults in WSNs

WSNs can range from a small handful to hundreds or even thousands of sensor nodes with various sensors monitoring diverse physical phenomena. Faulty sensor readings can lead to unnecessary shutdowns of plants or disruptions in monitored processes. While sensor technology advancements can reduce fault occurrences, they cannot be completely eliminated. Here, we present fault detection techniques, lists the common faults for sensors in WSNs, and identify mathematical models for such faults.

Sensor and actuator faults in complex, distributed electromechanical systems are well studied [33, 37, 48]. In [37], a systematic taxonomy of common sensor data faults that occur in deployed sensor networks and the detailed approaches commonly taken to model these faults are provided. These features include characteristics specific to the data, system, or environment pertaining to the system of interest. According to [37], the data features are usually statistical in nature. A confident diagnosis of any single fault may require more than one of these features to be modeled. The fault detection techniques most commonly used include mean and variance (determine expected behavior via regression models or correcting faulty sensor values), correlation (regression methods), gradient (rate of change), and spatial or temporal distance (determines if data is faulty). Table 4.2 shows common fault detection techniques that includes statistical, nearest neighbor, clustering, and classification methods.

Statistical methods create a model of healthy data and then classify any behavior that does not fit that model as a fault. For example, a statistical model of an outdoor temperature sensor may assume that temperature should be high during the day and low at night, and register a fault anytime that trend is not observed. In this case, a fault could either indicate a malfunction in the sensor network or an environmental irregularity, like a storm. Statistical methods have the advantage of justifiable mathematical models, but are not robust to phenomena that do not fit assumed distributions.

Fault classification methods take the opposite approach and instead of creating mathematical models for healthy data, they use mathematical models for faulty data. Incoming data from the WSN is analyzed for behavior that corresponds to the faulty behavior previously modeled. The main advantage of classification methods is that they allow users to specify an exact set of faults to be considered, while the main disadvantage is that they usually cannot detect faults not previously observed.

Nearest neighbor methods forego data models in favor of using a sensor-to-sensor comparison as a fault metric. These methods identify sets of sensors whose data should be similar (i.e., the spatially nearest sensors). Anytime these neighbors report significantly different readings from each other, a fault is registered. The main advantage of nearest neighbor methods is that they allow faults to be detected without a mathematical model for healthy or faulty data. The main disadvantage is that they require appropriate neighbors to be defined, and mistakes in deciding which sensors are related can result in poor data comparisons.

Clustering-based methods compare clusters of nodes rather than sensors. During a learning period, sets of nodes are grouped into clusters, which are then compared to each other to establish which clusters are measuring related data. During fault

Table 4.2 Common fault detection techniques

Method	References	Pros	Cons
Statistical	[3, 60]	Mathematically justified models	Fails if data does not fit assumed distribution
Classification	[35, 44]	Provides exact set of faults	Computationally expensive, requires proper kernel choice
Nearest neighbor	[5, 65]	No assumption on data distribution	Computationally expensive in large networks, dependent on input parameters
Clustering	[45]	No previous knowledge of data statistics is needed, adapts to new data	Cluster width mush be defined

detection, each cluster compares its data with its related clusters, and faults are registered whenever two related clusters are not measuring similar data. Clustering methods are less computationally expensive than nearest neighbor methods. The main disadvantage of clustering methods is the difficulty of appropriately defining clusters.

In [21], Principal Components Analysis (PCA) is used during a learning period to model healthy data behavior with the first four eigenvectors. Using these eigenvectors as a baseline for healthy behavior, any activity that falls outside of the healthy model is considered an outlier. In [35], a Support Vector Machine (SVM) is used to model healthy behavior and detect outliers in real time for a WSN. In [56], possible outliers are sorted and ranked periodically, speeding up the process of outlier discovery for a high dimensional dataset. Calibration methods focus on identifying and accounting for offsets and gains in a WSN. In [6], it is assumed that the phenomena being measured should have similar behavior everywhere and the characteristics of a small base set are used to model the characteristics of the entire network. In [2], it is assumed that the phenomena being measured have some linear correlation between a sensor and its nearest neighbor, and this information is used to calibrate the sensors in the WSN. The NASA Stennis Space Center has implemented a fault detection models for WSNs on a system that consists of multiple wireless sensor nodes, fault detection software, a server, and a client (Fig. 4.7). The system is intended for health monitoring applications of rocket test stands and widely distributed support systems, including pressurized gas lines, propellant delivery systems, and water coolant lines [16–19]. The main system components include a server, a Network Capable Application Processor (NCAP), Communication Module, a set of Transducer Interface Modules (TIMs), and wireless sensor nodes. The TIMs monitor sensors that measure physical phenomena, such as temperature, pressure,

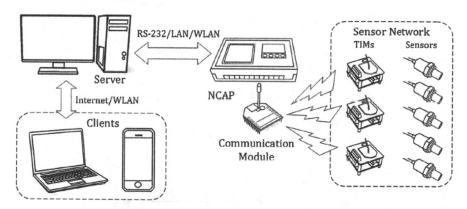

Fig. 4.7 Distributed wireless sensor network configuration that includes a server, a Network Capable Application Processor (NCAP), a communication module, a Transducer Interface Module (TIM) and wireless sensor nodes

and flow rate. They communicate measured data to the NCAP, which analyzes the data for faults. The NCAP records the data (as well as any detected faults) in a server where it can be accessed by a user.

4.5.3 Mathematical Models for Sensor Faults

We discuss here a model that is used in WSN fault detection and identification (Fig. 4.8) [54]. For a sensor node j, sensor readings from the node's multiple sensors are given by

$$Z_j(k) = \left[z_j^1(k) z_j^2(k) \ldots z_j^N(k) \right]^T, \tag{4.1}$$

where individual sensors on a node are identified as $z_j^i(k)$, $i = 1, 2, \ldots, N$. The model considers each node with an equal number of sensors, N, with output values at time instance k given by $z_j^i(k)$ (i-th sensor on j-th node at time k). A true value of a measured physical variable is $u_j^i(k)$ and a faulty measurement of the i-th sensor on the j-th node at time instance k is $\hat{z}_j^i(k)$.

There are two general categories for fault detection techniques: data-centric and system-centric [37]. Data-centric techniques focus on a single data stream to identify faults, while system-centric techniques consider the whole system to detect

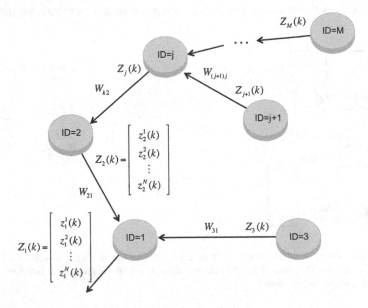

Fig. 4.8 Model of a WSN that is used in the fault analysis

faults. The two types of techniques need not be used exclusively. A computationally inexpensive data-centric technique may be used to identify abnormal behavior on a sensor stream, at which point a more expensive system-centric technique can be implemented that compares the sensor data to related neighboring nodes, verifying whether the abnormal data reflects an abnormal environment or a faulty sensor reading.

If a model for the behavior of a measured phenomenon is known, the model can be compared to the readings of a sensor, and the error between the two used for identification of faults. If a model is not known, then other techniques listed in Table 4.2 can be applied to identify faults.

4.5.3.1   Data-Centric Fault Models

For each type of fault, there are defining data characteristics, e.g. magnitude for outlier fault or variance for noise fault. One can identify a range for the specific data characteristic under normal operating conditions, and then determine a fault to have occured whenever the characteristic of interest falls outside of the expected range.

The energy constraints in WSNs limit communication and computation complexities, while one of the main hardware constraints is memory. For a large network, it is infeasible to store every data sample collected by a sensor. Therefore, online fault detection algorithms with a sliding data window are preferable ("one-pass algorithms", [22]). A sliding window considers the last W_{sl} data samples from a sensor data stream. Data samples are stored for the duration of the sliding window and then erased. This method is robust to changes in normal behavior of a sensor data.

A less robust and less costly method of defining data norms is to use a learning period. A standard assumption is that sensor data are healthy during the learning period. By analyzing a data stream over the learning period, *normal* behavior can be defined. The advantage is that a normal behavior has to be calculated only once. However, the method is not robust to data streams whose behavior changes with time. For example, the pressure in a pipe may be low during a learning period, but rise when gas is pumped into it. Using a sliding window allows algorithms to update the expected behavior of a data stream as its normal behavior changes, while still detecting true faults in the data stream.

Outlier Fault An outlier is a single data sample whose value is *significantly* (defined by the user) outside of the range defined by previous data samples, see Fig. 4.9. To quantify this range, one can define $\bar{Z}_j^i$, an upper bound of expected sensor data, and $\underline{Z}_j^i$, a lower bound of expected data.

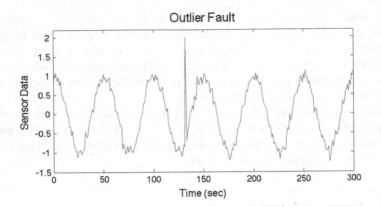

Fig. 4.9 Outlier occurring between 100 and 150 s

The upper and lower bounds $\overline{Z}_j^i$ and $\underline{Z}_j^i$ are given by

$$\overline{Z}_j^i(k) = \max\left\{z_j^i(p)\right\} + c_{\text{out}}\left|\max\left\{z_j^i(p)\right\} - \min\left\{z_j^i(p)\right\}\right|, \qquad (4.2)$$

$$\underline{Z}_j^i(k) = \min\left\{z_j^i(p)\right\} - c_{\text{out}}\left|\max\left\{z_j^i(p)\right\} - \min\left\{z_j^i(p)\right\}\right|, \qquad (4.3)$$

for $p = k_0, k_0 + 1, \ldots, k_0 + W_{\text{ln}} - 1$ where k_0 is the first data point in the learning period, W_{ln} is the number of data points in the learning period, and c_{out} is a positive constant that determines the outlier detection sensitivity. For instance, $c_{\text{out}} = 0.20$ allows the signal to vary 20 % of the range above and below data limits observed during the learning period. Increasing c_{out} lowers outlier detection sensitivity by allowing some faulty readings to go undetected, while reducing the false alarm rate.

When using a sliding window, $\overline{Z}_j^i(k)$ and $\underline{Z}_j^i(k)$ are recalculated periodically. Bounds can be recalculated every time new data is received. For a sliding window of a length W_{sl}, every time new data is received, the sliding window is updated, and the bounds become functions of the current data sample. The running index p in this case is $p = k - W_{\text{sl}}, k - W_{\text{sl}} + 1, \ldots, k - 1$. The benefit of using such bounds is that if the range of expected data changes, the bounds will adapt to them. However, frequent updating of the signal bounds increases the computational cost and is a natural tradeoff between the false alarm rate and computational cost. The condition for an outlier fault to be triggered is given by

$$z_j^i(k) > \overline{Z}_j^i(k) \quad \text{or} \quad z_j^i(k) < \underline{Z}_j^i(k). \qquad (4.4)$$

Spike Fault The term spike fault is used in reference of a small number, r_{s}, of data points that rise or fall more rapidly than the data during the healthy sensor behavior. Figure 4.10 shows a spike fault where data returns to normal behavior after the spike. The following parameters are used to identify a spike fault: $\overline{R}_j^i$, an upper

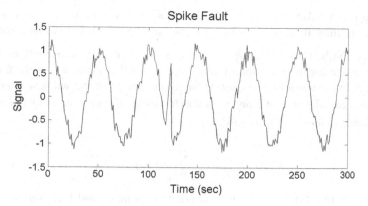

Fig. 4.10 Spike occurring between 100 and 150 s

bound on the gradient which may be ascribed to healthy data on the i-th sensor at the j-th node, and r_s, a number of successive samples required to have an unhealthy gradient before a spike occurs. The bound that is used in determining healthy behavior is based on a learning period or a sliding window

$$\overline{R}_j^i = (1 + c_{\mathrm{spk}}) \max \left| z_j^i(p) - z_j^i(p-1) \right| \tag{4.5}$$

where $p = k_0, k_0 + 1, \ldots, k_0 + W_{\mathrm{ln}} - 1$ in case of a fixed learning period and $p = k - r_s - W_{\mathrm{sl}}, k - W_{\mathrm{sl}} + 1, \ldots, k - r_s - 1$ in case of a sliding window, and c_{spk} is a positive constant that determines the detection sensitivity.

In case of using a sliding window, the gradient bound is a function of sample time k. The end of the sliding window should be set more than r_s samples away from the current data sample. The number of successive samples, r_s, required for a spike fault is influenced both by the phenomena being measured and the sampling rate. Some phenomena, such as the intensity of light, may be expected to produce signals with sharp gradients [37] while others are expected to change smoothly. For a slow-varying signal, a spike could indicate a fault in the sensor or that the sampling rate is too slow. A small r_s hastens the detection of spikes, which is crucial in time constrained applications. Conditions for spike detection are given by:

$$
\begin{aligned}
z_j^i(k) - z_j^i(k-1) &> \overline{R}_j^i & z_j^i(k) - z_j^i(k-1) &< -\overline{R}_j^i \\
z_j^i(k-1) - z_j^i(k-2) &> \overline{R}_j^i & \text{or} \quad z_j^i(k-1) - z_j^i(k-2) &< -\overline{R}_j^i \\
&\vdots & &\vdots \\
z_j^i(k - r_s + 1) - z_j^i(k - r_s) &> \overline{R}_j^i(k) & z_j^i(k - r_s + 1) - z_j^i(k - r_s) &< -\overline{R}_j^i(k)
\end{aligned}
$$

$$\tag{4.6}$$

There are two sets of conditions because a spike fault model requires r_s successive points to have either an abnormally large positive or negative gradients. Moreover, this eliminates false positive spike identification when high frequency noise is present, resulting in sensor readings having large gradients in random directions.

Variance Fault Variance fault is defined as a set of data whose values tend to differ from the mean by an abnormally large or small amount, where the threshold parameters are chosen by the user. Defining the length of the window over which variance is calculated as W_v, and the expected value of data over that window as $\bar{z}_j^i$, the variance is then given by [38]:

$$V_j^i(k) = \frac{1}{W_v - 1} \sum_{p=k-W_v}^{k-1} \left(z_j^i(p) - \bar{z}_j^i \right)^2. \tag{4.7}$$

There are two types of variance faults: low variance and high variance faults. A low variance fault (also called a stuck-at fault) occurs when the variance of a data stream is abnormally low, i.e., when a signal is stuck at a certain value. A high variance fault occurs when the variance is abnormally high. The model parameters include $\underline{V}_j^i$ and $\overline{V}_j^i$, a lower and an upper bound on signal variance under healthy conditions.

The variance fault has occurred if $V_j^i(k) > \overline{V}_j^i(k)$ or $V_j^i(k) < \underline{V}_j^i(k)$. Increasing the learning or sliding window size provides a larger sample size for determining the variance. If the phenomenon being measured is slow-varying throughout learning period, it will result in a more accurate representation of the steady state variance of the system. However, if the phenomenon is fast-varying over the learning period, signal changes will be reflected in the healthy variance parameters. A small window size will emphasize the role of variance as imperfections in the system, but will also increase false positives. Figures 4.11 and 4.12 show low variance and high variance faults, respectively.

There are several possible causes of variance faults. The term "noise" is often used to describe some high variance faults in the case of signals with particular

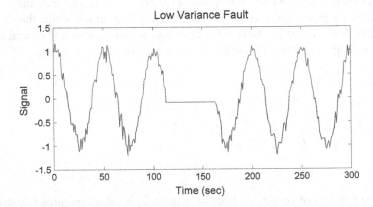

Fig. 4.11 Low variance fault occurring between 120 and 160 s

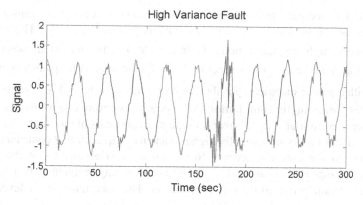

Fig. 4.12 High variance fault occurring between 150 and 200 s

mathematical attributes. Several types of noise exist, such as white noise, pink noise (equal power in bandwidths that are proportionally wide, see below), and violet noise (increases with frequency). Noise is usually classified by its behavior in the power spectrum and the fast Fourier transform (FFT) is often used to detect and isolate a noise fault [43].

High Frequency Noise Fault Noise classification and quantification is usually done in the frequency domain where the FFT is used to obtain the signal power spectrum, Fig. 4.13. Here, we describe a method for detecting undesired high frequency signals in the power spectrum and how to distinguish the presence of unexpected high frequency useful data and high frequency noise.

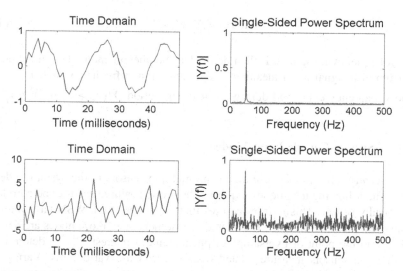

Fig. 4.13 Time domain and power spectrum for a sinusoid with small amount of white noise (*top*) and large amount of white noise (*bottom*)

Figure 4.13 indicates how the FFT can be used to detect noise in a sensor signal. Define the window of data on which the FFT is performed as W_{fft}. The variables that define a high frequency noise fault include $\overline{Y}_j^i$ (the maximum power level classified as noise in the power spectrum) and $\overline{v}_j^i$ (the highest frequency of a useful signal with a power spectrum contribution above the noise level $\overline{Y}_j^i$). To define what constitutes a meaningful contribution to the power spectrum, one needs to compute the power spectrum of a signal over a range of frequencies higher than $v = v_{\text{max}}$. It is assumed that v_{max} is significantly higher than any frequency that contributes more than noise to the power spectrum. In particular, it is assumed that the highest frequency component is at least W_{pow} higher than any frequency that makes a meaningful contribution to the power spectrum. The maximum power level which noise is expected to contribute, $\overline{Y}_j^i$, is then based on the power spectrum contributions of the last W_{pow} frequency components in the FFT:

$$\overline{Y}_j^i = \max\left\{Y_j^i(v)\right\} + c_{\text{fft}}\left|\max\left\{Y_j^i(v)\right\} - \min\left\{Y_j^i(v)\right\}\right|, v \in (v_{\text{max}} - W_{\text{pow}}, v_{\text{max}})$$

(4.8)

where c_{fft} is the parameter that is used to tune the sensitivity of the high frequency noise model. Any frequency component with a power spectrum contribution at or below $\overline{Y}_j^i$ is considered noise, and any frequency component which contributes a power level above $\overline{Y}_j^i$ is said to have a meaningful contribution to the data stream. The highest contributing frequency component under healthy conditions as the highest frequency v with a power spectrum contribution above $\overline{Y}_j^i$ is defined as:

$$\overline{v}_j^i = \max\left\{v_j^i | Y_j^i(v) > \overline{Y}_j^i\right\}.$$

(4.9)

A high frequency noise fault is then defined as the phenomenon wherein the power spectrum of a signal has a meaningful contribution at a frequency higher than the highest frequency expressed during the learning period, $Y_j^i(v) > \overline{Y}_j^i$ for $v > \overline{v}_j^i$.

4.5.3.2 System-Centric Fault Models

System-centric fault models use data from multiple sensors in the system to detect faults. The following techniques require at least one healthy sensor to be correlated to the faulty sensor. This in turn requires that the nodes in the WSN have multiple sensors that measure the same (or related) phenomena, i.e., those nodes are densely deployed to the point of oversampling phenomena of interest [2]. To determine if two or more sensors measure related data, one can use variograms. Variograms quantify the correlation of a phenomenon at one point in the system with a phenomenon at another point in the system. A variogram $\gamma^i_{j,l}$ between sensors j and l is defined as:

$$\gamma^i_{j,l}(k) = \frac{1}{2W_{\text{vgm}}} \sum_{p=k}^{k-W_{\text{vgm}}+1} \left(z^i_j(p) - z^i_l(p)\right)^2, \tag{4.10}$$

where W_{vgm} is the window of data where the variogram is being calculated. In the event of spatial correlation among sensors, the variogram is a function of radius r around sensor j:

$$\gamma^i_j(k) = \frac{1}{2W_{\text{vgm}}|\Omega_j(r)|} \sum_{q \in \Omega_j(r)} \sum_{p=k}^{k-W_{\text{vgm}}+1} \left(z^i_j(p) - z^i_q(p)\right)^2, \tag{4.11}$$

where r is the radius around sensor j, $\Omega_j(r)$ is the set of all neighbors of sensor j within radius r, and $|\Omega|$ is the cardinality of the set Ω. A small variogram implies a high correlation among the sensors. If the variogram between a sensor and its neighbors is smaller than some threshold Γ, then the sensor reading is related to its neighbors.

Offset Fault An offset fault occurs when sensor data values are *offset* from the true phenomenon being measured by a constant value:

$$\hat{z}^i_j(k) = f\left(u^i_j(k)\right) + \beta_0, \tag{4.12}$$

where the function f represents a nonlinear sensor model, $u^i_j(k)$ is a true value being measured on the i-th sensor at the j-th node, and β_0 is the offset. Figure 4.14 shows an example of an offset fault. To determine the offset value, one usually requires

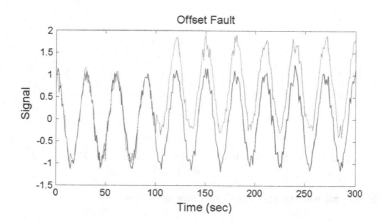

Fig. 4.14 A signal with a constant offset fault

either a ground truth value of the sensor readings or a precise sensor model f, see
[16]. In case of a densely distributed sensor network, an offset fault can be modeled
as a constant difference between the readings of a sensor and the readings of its
related neighbors.

Let us consider the difference between a sensor's readings and the average of the
readings of its related neighbors:

$$\Delta_j^i(k) = z_j^i(k) - \frac{1}{|\Omega_j|} \sum_{q \in \Omega_j} z_q^i(k). \tag{4.13}$$

A constant offset is modeled as a slow-varying difference between a sensor's
measurements and the true value being measured. To distinguish between a constant offset and a time-changing offset (which may be considered a drift fault, see
below) the variance, $\text{var}\left(\Delta_j^i(k)\right)$, over a window of a certain size is observed. The
conditions for an offset fault are that the difference between a sensor's readings and
the readings of its related neighbors be above the threshold value and that the
variance of the difference is sufficiently small, i.e., $\Delta_j^i(k) \geq c_{\text{oft}}$ and
$\text{var}\left(\Delta_j^i(k)\right) \leq \beta_{\text{oft}}$, where c_{oft} and β_{oft} are model parameters.

Gain Fault A gain fault occurs when the sensor data for a certain measured
phenomenon differ from the healthy sensor data by a constant ratio; see Fig. 4.15.
In the event of a gain fault, it is expected that the gain of a sensor's readings
compared to the average of its neighbor's readings, η_j^i, to be significantly lower or
higher than 1. This ratio can be found as

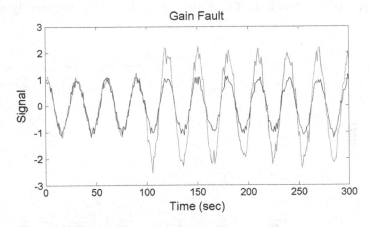

Fig. 4.15 A signal with the gain fault (higher gain in this case)

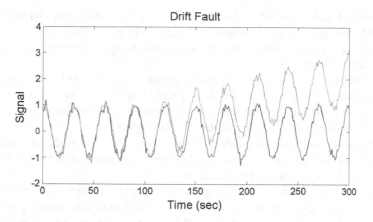

Fig. 4.16 Drift fault when the offset between the signals is steadily increasing

$$\eta_j^i(k) = \frac{z_j^i(k)}{\frac{1}{|\Omega_j|} \sum_{q \in \Omega_j} z_q^i(k)}.$$ (4.14)

To quantify variation of the sensor's gain, we consider the variance $\mathrm{var}\left(\eta_j^i(k)\right)$ over the window of data. A gain fault has occurred when $\left|\eta_j^i(k) - 1\right| \geq c_{gn}$ and $\mathrm{var}\left(\eta_j^i(k)\right) \leq \beta_{gn}$, where c_{gn} and β_{gn} are small, adjustable parameters.

Drift Fault A drift fault occurs when sensor readings drift away from the real values by an amount that increases with time, Fig. 4.16.

Given $\Delta_j^i(k)$ in (4.13), a drift fault model represents a steady increase in time, i.e., an offset increment $\lambda_j^i(k, L) = \Delta_j^i(k) - \Delta_j^i(k - L)$ is a function of time increment L:

$$\lambda_j^i(k, L) = \lambda_j^i(L) = c_{dft}.$$ (4.15)

If c_{dft} varies with time, then (4.15) models a drift that is increasing according to the integral of c_{dft}.

Questions and Exercises

1. What are the advantages of random key pre-distribution (RKP) relative to centralized key distribution in WSNs?
2. Briefly describe the shortcomings of RKP and the general philosophy of operation of some of the schemes that have been designed to address these shortcomings.

3. Briefly describe the steps through which a keyed checksum can be used to verify the integrity of messages exchanged between WSN nodes.
4. Under what conditions would it be feasible to apply asymmetric key cryptography in WSNs?
5. Use MATLAB to simulate an RKP scheme in which 1000 nodes create key rings containing m keys drawn from a pool of 50 keys for $m = 2, 5, 10$ and 20. Comment on the connectivity of the network for the different values of m. Assume the keys are the numbers 1 through 50.
6. What is the difference between a wormhole attack and a black hole attack? In your opinion which of these attacks would be more difficult to detect in practice? Give a reason for your answer.
7. Assume that sensor data follow sinusoidal function in time with amplitude of 10. Give an example of an outlier sensor fault and how would you set the threshold parameters in such a case? Does the sampling rate affects the fault model and how?
8. What would be the scenario where the spike fault would overlap with the outlier fault? Can this be prevented?
9. How does the variance fault differ from the high frequency noise fault? Which class set is a superset of the other?
10. Use MATLAB to simulate a high frequency noise fault for a simple data set you create. Then tune fault parameters such that the fault can be effectively detected.
11. Suppose that a sensor is producing sinusoidal type of data over time $z_1(t) = 5 \sin(wt)$. At a certain moment there is a gain fault of value 2. How would a faulty data look like and what other faults could be triggered at that moment? Would any other fault be triggered at steady state value of sensor data, i.e., after fault has already occurred?
12. Consider a real life scenario of a sensor with saturation limits at the output. If the drift fault occurs, after some time, what would such fault look like? Any fault detection after long time period will identify what kind of sensor problem?

References

1. F. Anjum, "Location dependent key management using random key-predistribution in sensor networks," in *Proc. 5th ACM Workshop on Wireless Security*, Los Angeles, California, 2006.
2. L. Balzano and R. Nowak, "Blind calibration of networks of sensors: Theory and algorithms," in *Networked Sensing Information and Control*, V. Saligrama, Ed. Springer US, 2008, pp. 9–37.
3. L. Bettencourt, A. Hagberg, and L. Larkey, "Separating the wheat from the chaff: Practical anomaly detection schemes in ecological applications of distributed sensor networks," in *Proceedings of the IEEE International Conference on Distributed Computing in Sensor Systems*, 2007.
4. R. Beyah, J. McNair, and C. Corbett, Ed., *Security in Ad Hoc and Sensor Networks*, World Scientific Publishing Co, Singapore, 2010.

5. J. Branch, B. Szymanski, C. Giannella, and R. Wolff, "In-network outlier detection in wireless sensor networks," in *Proceedings of the IEEE Conference on Distributed Computing Systems*, 2006.

6. V. Bychkovskiy, S. Megerian, D. Estrin, and M. Potkonjak, "A collaborative approach to in-place sensor calibration," in *Proceedings of the Second International Workshop on Information Processing in Sensor Networks* (IPSN, 2003, pp. 301–316.

7. A. Cerpa, J. Elson, D. Estrin, L. Girod, M. Hamilton, and J. Zhao, "Habitat monitoring: application driver for wireless communications technology," *SIGCOMM Comput. Commun. Rev.*, vol. 31, pp. 20–41, 2001.

8. H. Chan, A. Perrig, and D. Song, "Random Key Predistribution Schemes for Sensor Networks," in *Proc. 2003 IEEE Symposium on Security and Privacy*, 2003.

9. H. Chan, A. Perrig, and D. Song, "Key Distribution Techniques for Sensor Networks," *Wireless Sensor Networks*, C. S. Raghavendra, K. Sivalingam, and T. Znati, Eds., ed: Springer US, 2004, pp. 277–303.

10. S. Chessa and P. Santi, "Crash Fault Identification in Wireless Sensor Networks," *Computer Communications*, vol. 25, no. 14, pp. 1273–1282, 2002.

11. W. Chonggang, K. Sohraby, L. Bo, M. Daneshmand, and H. Yueming, "A survey of transport protocols for wireless sensor networks," IEEE Network, vol. 20, pp. 34–40, 2006.

12. W. Du, J. Deng, Y.S. Han, and P.K. Varshney, "A pairwise key pre-distribution scheme for wireless sensor networks," in *Proc. the 10th ACM Conference on Computer and Communications security*, Washington D.C., USA, 2003.

13. E. Elnahrawy and B. Nath, "Clearing and Querying Noisy Sensors," *Proc. Workshop on Sensor Networks and Applications (WSNA)*, 2003.

14. J. Elson, L. Girod, and D. Estrin, "Fine-grained network time synchronization using reference broadcasts," *SIGOPS Oper. Syst. Rev.*, vol. 36, pp. 147–163, 2002.

15. L. Eschenauer and V.D. Gligor, "A key-management scheme for distributed sensor networks," in *Proc. 9th ACM Conference on Computer and Communications Security*, Washington, DC, USA, 2002.

16. F. Figueroa, J. Schmalzel, J. Morris, M. Turowski, and R. Franzl, "Integrated system health management: Pilot operational implementation in a rocket engine test stand," in *AIAA Infotech@Aerospace 2010*, Atlanta, GA, April 2010.

17. F. Figueroa, J. Schmalzel, R. Aguilar, M. Shwabacher, and J. Morris, "Integrated system health management (ISHM) for test stand and j-2x engine: Core implementation," in *44ᵗʰ AIAA/ASME/SAE/ASEE Joint Propulsion Conference and Exhibit*, Hartford, CT, July 2008.

18. F. Figueroa, J. Schmalzel, M. Walker, M. Venkatesh, R. Kapadia, J. Morris, M. Turowski, and H. Smith, "Integrated system health management: Foundational concepts, approach, and implementation," *NASA Stennis Space Center, Tech. Rep.*, April 2009.

19. F. Figueroa, J. Schmalzel, R. Aguilar, M. Shwabacher, and J. Morris, "Tutorial integrated systems health management (ISHM)," in *NASA/ESA Conference on Adaptive Hardware and Systems (AHS-2011)*, San Diego, CA, June 2011.

20. G. Gaubatz, J.-P. Kaps, and B. Sunar, "Public Key Cryptography in Sensor Networks—Revisited," *Security in Ad-hoc and Sensor Networks*, vol. 3313, C. Castelluccia, H. Hartenstein, C. Paar, and D. Westhoff, Eds., ed: Springer Berlin Heidelberg, 2005, pp. 2–18.

21. J. Gupchup, A. Terzis, R.C. Burns, and A.S. Szalay, "Model-based event detection in wireless sensor networks," *Computing Research Repository* arXiv:0901.3923, 2009.

22. C. Han, L. Xu, and G. He, "Mining recent frequent itemsets in sliding windows over data streams," *Computing and Informatics*, vol. 27, pp. 315–339, 2008.

23. C. Hartung, J. Balasalle, and R. Han, "Node Compromise in Sensor Networks: The Need for Secure Systems," University of Colorado at Boulder Technical Report CU-CS-990-052005.

24. J. Hwang and Y. Kim, "Revisiting random key pre-distribution schemes for wireless sensor networks," in *Proc. 2nd ACM Workshop on Security of Ad Hoc and Sensor Networks*, Washington DC, USA, 2004.

25. C. Intanagonwiwat, R. Govindan, and D. Estrin, "Directed diffusion: a scalable and robust communication paradigm for sensor networks," in *Proc. 6th Annual International Conference on Mobile Computing and Networking*, Boston, Massachusetts, USA, 2000.

26. C. Karlof and D. Wagner, "Secure Routing in Wireless Sensor Networks: Attacks and Countermeasures," in *Proc. First IEEE International Workshop on Sensor Network Protocols and Applications*, 2002, pp. 113–127.

27. F. Koushanfar, M. Potkonjak, and A. Sangiovanni-Vincentelli, "Fault tolerance techniques for wireless ad hoc sensor networks," *Proceedings of IEEE Sensors*, 2002, pp. 1491–1496 vol. 2.

28. F. Koushanfar, M. Potkonjak, and A. Sangiovanni-Vincentelli, "Fault Tolerance Techniques for Wireless Ad Hoc Sensor Networks," *IEEE Sensors*, vol. 2, pp. 1491–1496, 2002.

29. B. Krishnamachari and S. Iyengar, "Distributed Bayesian algorithms for fault-tolerant event region detection in wireless sensor networks," *IEEE Transactions on Computers*, vol. 53, pp. 241–250, 2004.

30. B. Krishnamachari and S. Iyengar, "Distributed Bayesian Algorithms for Fault-Tolerant Event Region Detection in Wireless Sensor Networks," *IEEE Transactions on Computers*, vol. 53, no. 3, 2004.

31. J. Lee and D. Stinson, "Deterministic Key Predistribution Schemes for Distributed Sensor Networks," *Selected Areas in Cryptography*, vol. 3357, H. Handschuh and M. A. Hasan, Eds., ed: Springer Berlin Heidelberg, 2005, pp. 294–307.

32. D. Liu and P. Ning, "Establishing pairwise keys in distributed sensor networks," in *Proc. 10th ACM Conference on Computer and Communications Security*, Washington D.C., USA, 2003.

33. X. Luo, M. Dong, and Y. Huang, "On distributed fault-tolerant detection in wireless sensor networks," *IEEE Transaction on Computers*, vol. 55, no. 1, pp. 58–70, 2006.

34. K. Marzullo, "Tolerating Failures of Continuous Valued Sensors," *ACM Transactions on Computer Systems*, vol. 8, pp. 284–304, 1990.

35. M.S. Mohamed and T. Kavitha, "Outlier detection using support vector machine in wireless sensor network real time data," *IEEE Journal of Soft Computing and Engineering*, vol. 1, no. 2, pp. 68–72, May 2011.

36. J. Newsome, E. Shi, D. Song, and A. Perrig, "The Sybil attack in sensor networks: analysis and defenses," *Proc. Third International Symposium on Information Processing in Sensor Networks*, 2004, pp. 259–268.

37. K. Ni, N. Ramanathan, M.N.H. Chehade, L. Balzano, S. Nair, S. Zahedi, G. Pottie, M. Hansen, M. Srivastava, and E. Kohler, "Sensor network data fault types," *ACM*, vol. 5, no. 3, pp. 1–29, August 2009.

38. R.L. Ott and M.T. Longnecker, *An Introduction to Statistical Methods and Data Analysis*, Cengage Learning, 6th Edition, 2008.

39. A.S.K. Pathan, H.-W. Lee, and C.S. Hong, "Security in wireless sensor networks: Issues and challenges," *Proc. 8th International Conference on Advanced Communication Technology, ICACT 2006*, 2006.

40. A. Perrig, R. Szewczyk, J.D. Tygar, V. Wen, and D. Culler, "SPINS: Security Protocols for Sensor Networks," *Wireless Networks*, vol. 8, pp. 521–534, 2002.

41. R.D. Pietro, L.V. Mancini, and A. Mei, "Random key-assignment for secure Wireless Sensor Networks," in *Proc. 1st ACM Workshop on Security of Ad Hoc and Sensor Networks*, Fairfax, Virginia, 2003.

42. K. Piotrowski, P. Langendoerfer, and S. Peter, "How public key cryptography influences wireless sensor node lifetime," in *Proc. 4th ACM Workshop on Security of Ad Hoc and Sensor Networks*, Alexandria, Virginia, USA, 2006.

43. J. Proakis and D. Manolakis, *Digital Signal Processing*, Prentice Hall, 2006.

44. S. Rajasegarar, C. Leckie, M. Palaniswami, and J. Bezdek, "Quarter sphere based distributed anomaly detection in wireless sensor networks," in *Proceedings of the IEEE International Conference on Communications*, 2007.

45. S. Rajasegarar, C. Leckie, M. Palaniswami, and J. Bezdek, "Distributed anomaly detection in wireless sensor networks," in *Proceedings of the IEEE International Conference on Communication Systems*, 2006.

46. D.R. Raymond and S.F. Midkiff, "Denial-of-Service in Wireless Sensor Networks: Attacks and Defenses," *IEEE Pervasive Computing*, vol. 7, pp. 74–81, 2008.

47. "Coremicro reconfigurable embedded smart sensor node (CRE-SSN) product brochure, url: http://americangnc.com/images/cre-ssn brochure.pdf, accessed Feb. 2015." American GNC Corporation, Tech. Rep., 2012.

48. M. Russell, G. Lecakes, S. Mandayam, and S. Jensen, "The intelligent valve: A diagnostic framework for integrated system-health management of a rocket-engine test stand," *IEEE Transactions on Instrumentation and Measurement*, vol. 60, no. 4, pp. 1489–1497, April 2011.

49. Y. Sankarasubramaniam, Ö.B. Akan, and I.F. Akyildiz, "ERST: Event-to-Sink Reliable Transport in Wireless Networks," *ACM MobiHoc'03*, 2003.

50. M. Shaneck, K. Mahadevan, V. Kher, and Y. Kim, "Remote Software-Based Attestation for Wireless Sensors," in *Security and Privacy in Ad-hoc and Sensor Networks*, vol. 3813, R. Molva, G. Tsudik, and D. Westhoff, Eds., ed: Springer Berlin Heidelberg, 2005, pp. 27–41.

51. B. Shen, "Application of Error Correction Codes in Wireless Sensor Networks," Master of Science (MS), The University of Maine, 2007.

52. E. Shi and A. Perrig, "Designing secure sensor networks," *Wireless Communications, IEEE*, vol. 11, pp. 38–43, 2004.

53. M. Srivastava, "Energy Aware Wireless Sensor and Actuator Networks," CENS, 2005.

54. N. Vosburg, R. Selmic, S. Oonk, and F. Maldonado, "Intelligent distributed and ubiquitous health management system: Data storage and processing," in *AIAA Infotech@Aerospace*, Boston, MA, August 2013.

55. C.Y. Wan, A.T. Campbell, and L. Krishnamurthy, "PSFQ: A Reliable Transport Protocol for Wireless Sensor Networks," *ACM International Workshop on Wireless Sensor Networks and Applications (WSNA '02)*, 2002.

56. Y. Wang, S. Parthasarathy, and S. Tatikonda, "Locality sensitive outlier detection: A ranking driven approach," in *Proc. IEEE 27th International Conference on Data Engineering (ICDE)*, 2011, pp. 410–421.

57. T.-Y. Wang, L.-Y. Chang, D.-R. Duh, and J.-Y. Wu, "Fault-tolerant decision fusion via collaborative sensor fault detection in wireless sensor networks," *IEEE Transactions on Wireless Communications*, vol. 7, no. 2, pp. 756–768, 2008.

58. R. Watro, D. Kong, S.-F. Cuti, C. Gardiner, C. Lynn, and P. Kruus, "TinyPK: securing sensor networks with public key technology," in *Proc. 2nd ACM Workshop on Security of Ad Hoc and Sensor Networks*, Washington DC, USA, 2004.

59. X. Wenyuan, M. Ke, W. Trappe, and Y. Zhang, "Jamming sensor networks: attack and defense strategies," *IEEE Network*, vol. 20, pp. 41–47, 2006.

60. W. Wu, X. Cheng, M. Ding, K. Xing, F. Liu, and P. Deng, "Outlying and boundary data detection in sensor networks," *IEEE Transactions on Knowledge and Data Engineering*, vol. 19, no. 8, pp. 1145–1157, 2011.

61. K. Xing, S. Srinivasan, M.M. Rivera, J. Li, and X. Cheng, "Attacks and Countermeasures in Sensor Networks: A Survey," in Network Security, S. C. H. Huang, D. MacCallum, and D.-Z. Du, Eds., ed: Springer US, 2010, pp. 251–272.

62. W. Yong, G. Attebury, and B. Ramamurthy, "A survey of security issues in wireless sensor networks," *Communications Surveys & Tutorials, IEEE*, vol. 8, pp. 2–23, 2006.

63. Z. Yun and F. Yuguang, "Scalable and deterministic key agreement for large scale networks," *IEEE Transactions on Wireless Communications*, vol. 6, pp. 4366–4373, 2007.

64. Z. Yun, F. Yuguang, and Z. Yanchao, "Securing wireless sensor networks: a survey," *Communications Surveys & Tutorials, IEEE*, vol. 10, pp. 6–28, 2008.

65. K. Zhang, S. Shi, H. Gao, and J. Li, "Unsupervised outlier detection in sensor networks using aggregation tree," in *Proceedings of the Advanced Data Mining and Applications*, 2007.

66. S. Zoican, "Frequency hopping spread spectrum technique for wireless communication systems," in *Proc. 5th International Symposium on Spread Spectrum Techniques and Applications*, pp. 338–341 vol. 1, 1998.

Chapter 5
Coverage and Connectivity

A major advantage of wireless sensor networks (WSNs) over wired networks is the potential for ad hoc deployment of the network. If the monitoring of a dangerous environment is required, then one may not be able to deploy a wired network. However, a WSN may be deployed in even the most inhospitable of domains, and the sensor data can then be gathered and monitored from a remote location. A likely ad hoc deployment method for distributing the sensor nodes in such a hazardous terrain is to airdrop them over the region of interest. In a chemical monitoring application, for instance, wireless sensor nodes may be airdropped over an area which could be unsafe to manually deploy the nodes and then self-organize them into a network for this monitoring task.

The problem with this deployment method, as well as other random deployment methods often used with WSNs, is that one has no full control over where the nodes will be located. Consider, as an example, the application of battlefield surveillance in which a large number of sensor nodes, designed to detect and locate the presence of enemy snipers [30], are airdropped over the battlefield. Nodes may use acoustic sensing to detect gunshots fired by snipers and approximate the location of a shooter by aggregating the time of arrival (ToA) data from several sensors. Crucial to this application, as well as many others, is the often implicit assumption that the deployment of sensor nodes is complete such that any event occurring within the sensor field will be detected. However, due to the randomness of a deployment there is no guarantee of complete coverage, raising questions on how to verify the absence of holes in the sensor coverage within the region of interest, and how to locate such holes in coverage if they do exist.

The coverage of the area of interest is one of the Qualify of Service (QoS) metrics in WSNs [18] as it describes how well the sensing field is sensed or covered by the sensor nodes. A point in the region of interest is considered to be covered if one or more sensors can measure the phenomena of interest at that point. If every point in the sensing field is covered by at least one sensor node, we say that the coverage is complete, and such coverage has degree of one. Sometimes specific

© Springer International Publishing AG 2016
R.R. Selmic et al., *Wireless Sensor Networks*,
DOI 10.1007/978-3-319-46769-6_5

applications require higher fault redundancy or sensor network measurement accuracy, resulting in higher degree of coverage where multiple sensor nodes cover point(s) in the sensing field.

Related to coverage is connectivity—another important characteristic of WSNs. As described later in the chapter, sensor networks can be modeled using graphs where nodes are equivalent to vertices and communication links are represented by corresponding edges. If the equivalent graph is connected then we consider the sensor network to be connected. The graph (or sensor network) is connected if there is an edge path (communication path) between any two vertices (nodes) in the graph (network). Otherwise, the graph (network) is disconnected and some nodes are not able to communicate with the rest of the network. Disconnected network nodes are basically useless for the network since the information on those nodes is not accessible by the network anymore. Similarly, as with the coverage in the sensor network, connectivity can have higher degrees that allows for more robust networks where removal or failure of some nodes does not cause the network to become disconnected.

Notions of coverage and connectivity are, in most practical applications, not independent of each other. Most applications have strict requirements for both of these network characteristics. It is often required to provide a certain level (quality) of coverage, while maintaining the network connectivity. This translates into constrained optimization problems, discussed later in the chapter, where type of coverage determines optimization cost function and connectivity relates to constraints. Similarly, it might be required to deploy a robust network that is fault-tolerant with a higher degree of connectivity, while the sensing area is still fully covered.

In this chapter, we review important notation and results related to sensor network coverage and connectivity and tools that are commonly used in modeling of sensor networks coverage and connectivity. We present basic graph notations that are used in WSN modeling including tools used in coverage modeling and provide few examples of optimal coverage under sensor network connectivity constraints.

5.1 Modeling Sensor Networks Using Graphs

In this section, we cover modeling of wireless sensor network using certain mathematical constructs known as graphs and simplicial complexes. This mapping of the sensor network allows users to perform mathematical and computational analysis of the network's topology, to analyze network coverage and connectivity, and develop algorithms that deal with those issues. Such modeling is, in general, applicable to both networks with known coordinates and coordinate-free networks.

5.1.1 Communication Graphs

A *graph*, defined as $G = (V, E)$ describes a set V of vertices and a set E of edges that connect the vertices. In most cases in WSN applications, it is presumed that all communication links between sensor nodes are bidirectional. This means that if node A is within the communication range of node B, then the reverse is also true. For this reason, this discussion of graphs may be restricted to deal only with *undirected simple graphs*. A graph is called a simple graph if it does not contain multiple edges between pairs of vertices or self-loops, which connect a vertex to itself, and it is called undirected if all edges are bidirectional. For simplicity, the term graph is often used to refer to such graphs. Figure 5.1 illustrates the graph $G = (V, E)$ with vertex set $V = \{1, 2, 3, 4\}$, and edge set $E = \{12, 13, 14, 24, 34\}$. Note also that the edge set for an undirected graph may consist of unordered pairs of vertices such that the edge 12 is equivalent to the edge 21.

Here, we define terms such as vertex degrees and order of a graph [54] that are used in WSNs modeling.

Definition 5.1 The degree of vertex v in a graph G, $d(v)$ is the number of edges incident to v. If the maximum degree in a graph is equal to the minimum degree in the graph, then the graph is called regular graph. The graph is called k-regular if the common degree is k.

For example, the degree of vertex 1 in the graph in Fig. 5.1 is 3, while the degree of vertex 3 is 2.

Definition 5.2 The order of a graph G, $n(G)$, is the number of vertices in graph G. The size of a graph G, $e(G)$, is the number of edges in graph G.

Definition 5.3 [54] A graph G is connected if it has a u, v-path whenever $u, v \in V(G)$; otherwise, G is disconnected. If graph G has a u, v-path, then vertex u is connected to vertex v in G.

Directed Graphs Communication links in sensor networks does not need to be symmetric. If the node A can hear the node B, it does not necessary mean that the node A can hear the node B. This can happen due to several reasons including node

Fig. 5.1 Example of a simple graph with four vertices and five edges

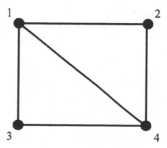

failure, different energy levels at the node that force radio to operate at different power levels, etc. A model that represents such a network is called directed graph.

Definition 5.4 [54] A directed graph G is a triple consisting of a vertex set $V(G)$, an edge set $E(G)$, and a function that assigns to each edge an ordered pair of vertices. The first vertex of the ordered pair is the tail of the edge, and the second is the head.

An example of a directed graph is given in Fig. 5.2, where there is an edge from 3 to 1, from 3 to 2, and from 3 to 4. There is also an edge from 1 to 2, and from 4 to 2.

Definition 5.5 An adjacency matrix $A(G)$ of a directed graph G is a matrix that at the position i, j has the number of edges from v_i to v_j.

For example, an adjacency matrix for the graph in Fig. 5.2 is given by

$$A(G) = \begin{matrix} & \begin{matrix} 1 & 2 & 3 & 4 \end{matrix} \\ \begin{matrix} 1 \\ 2 \\ 3 \\ 4 \end{matrix} & \begin{pmatrix} 0 & 1 & 0 & 0 \\ 0 & 0 & 0 & 0 \\ 1 & 1 & 0 & 1 \\ 0 & 1 & 0 & 0 \end{pmatrix} \end{matrix}. \tag{5.1}$$

If the graph is weighted, then for each pair of vertices u, v, there is a assigned weight a_{uv}, [34]. We consider a graph with weights that are real, satisfying $a_{uv} = a_{vu}$, $a_{uv} \neq 0$ if and only if u and v are adjacent vertices, and $a_{uv} \geq 0$. In case of weighted graphs, the adjacency matrix is $A(G) = [a_{uv}]_{u,v \in V(G)}$ and the degree of a vertex v is equal to a sum of all weights that are adjoining to the vertex v, i.e., $d(v) = \sum_u a_{uv}$.

Definition 5.6 A complement of a graph G is a graph $\bar{G}$ that has the same set of vertices and a complement set of edges, i.e., there is an edge between any two vertices in $\bar{G}$ if there is no edge between the same vertices in G.

Note that together, a graph G and its complement $\bar{G}$ form a complete graph with all possible edges, see Fig. 5.3.

WSNs are commonly mapped to corresponding graphs that represent connectivity of the nodes and can be used to solve certain problems. The set of sensor

Fig. 5.2 Example of a directed graph

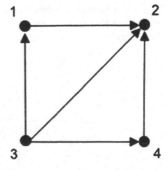

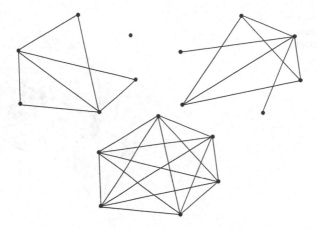

Fig. 5.3 A graph G, its compliment $\bar{G}$ and a complete graph $G + \bar{G} = K_n$

nodes N in the network becomes the set of vertices in the graph. Two nodes are then connected with a graph edge if there is a direct communication link between them. By direct communication link, we mean only those nodes that can communicate directly without multi-hopping. This is illustrated in Fig. 5.4 below where a sensor network is represented as a graph.

Assuming that communication range between sensor nodes is r_c and that the communication model is binary (when inside the communication range nodes can communicate, outside the range they cannot), then any two nodes whose Euclidean distance is less than r_c have a corresponding graph edge in the equivalent communication graph. Equivalently, if two nodes lie in each other's communication

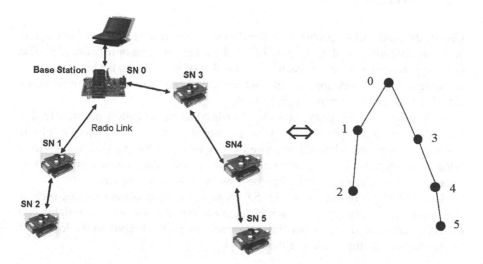

Fig. 5.4 Mapping a sensor network to a graph

Fig. 5.5 A sensor network
and its communication graph
—a unit disk graph

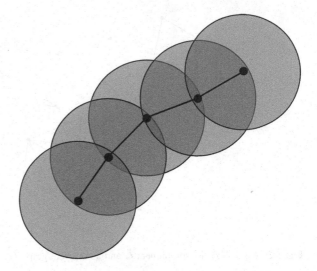

disk, then there is an edge between them. Therefore, the sensor network is modeled using a specific type of graph known as a *unit disk graph*, which admits a graph edge between two vertices if their Euclidean distance is less than a fixed threshold. For communication networks, this graph is commonly referred to as the *communication graph* of the network. An example of a communication graph for a simple network is given in Fig. 5.5. The disks around each node represent the communication range r_c of the nodes.

5.2 Coverage

One of the fundamental problems in the field of WSNs is the coverage. Coverage is an important indicator of the Quality of Service (QoS) in a sensor network [33]. The coverage problem can be approached in a number of ways due to the broad range of possible sensor network applications, but the central goal is to determine how well a set of sensor nodes monitors a given area.

The sensor coverage problem is closely related to the art gallery problem [37], a well-known visibility problem from computational geometry. The objective of this problem, which was originally proposed in August 1973 by mathematician Victor Klee, is to determine a minimum number of observers needed for surveillance of an art gallery such that every point can be seen by at least one observer.

This problem is illustrated in Fig. 5.6. Note that for any convex region, such as that to the left, only one observer is required. For the region to the right, two observers are needed to monitor the entire area since one observer to the lower left corner cannot see the portion at the top right.

Fig. 5.6 The art gallery
problem example

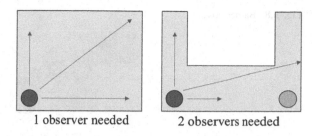

1 observer needed   2 observers needed

This problem is shown to be NP-hard in [31]. Reference [7] proves that $\lfloor n/3 \rfloor$ observers are always sufficient and the solution is sometimes optimal in 2D space, where n is the number of sides in the polygonal region being monitored, and the floor function $\lfloor x \rfloor$ is defined as the largest integer not greater than x (for example, $\lfloor 3.2 \rfloor = 3$, $\lfloor -3.2 \rfloor = -4$). This result is commonly referred to as Chvatal's art gallery theorem. A linear-time algorithm for locating these $\lfloor n/3 \rfloor$ observers is given in [27]. An approximation algorithm for the 3D case is proposed in [32] which is within $O(\log n)$ of the optimal solution.

In [16], coverage is classified into three main categories: blanket coverage, barrier coverage, and sweep coverage. The goal of blanket coverage, as name suggests, is to maximize the coverage of area of interest. This is the common objective for a majority of WSN monitoring applications. Barrier coverage attempts to minimize the probability of an intruder penetrating the barrier and is important security feature in WSN applications. An example of barrier coverage is border patrol, where a nation attempts to keep people from crossing its borders illegally [3, 29]. Finally, sweep coverage is the equivalent of a barrier moving across the area of interest where a set of sensors sweeps specific area. This incorporates aspects of both blanket coverage and barrier coverage. These three types of coverage are shown below in Figs. 5.7, 5.8, and 5.9, respectively.

A further issue to consider with coverage problems is point coverage versus area coverage. Either the sensor nodes can focus coverage on certain areas or targets, referred to as point coverage, or they can be spread out in an attempt to cover an

Fig. 5.7 Blanket coverage

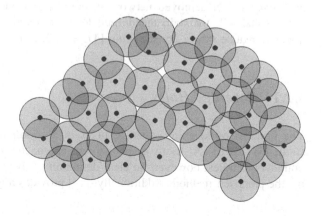

Fig. 5.8 Barrier coverage

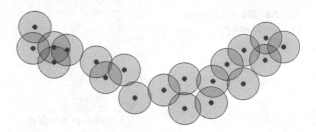

Fig. 5.9 Sweep coverage

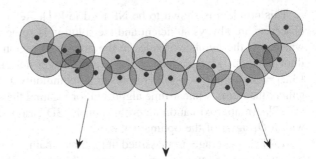

entire area, referred to as area coverage. Typically, in most common WSN applications, area coverage is used. However, there are some situations, such as in tracking applications, where point coverage may be more appropriate.

The topic of sensing coverage is often approached by studying how to position the sensors such that the area of interest is sufficiently monitored. A non-complete coverage in WSNs (coverage with holes) allows for intrusion into the system and presents a security issue where a WSN needs to deal with intrusion detection in some other way [2, 42]. Several authors have considered the problem of optimally positioning sensor such that a region is maximally covered [6, 23, 26, 58]. The central idea is to spread out the nodes as much as possible to improve sensing coverage over a region of interest. Other results have focused on finding a minimal number of sensors needed to monitor an area [1, 38]. This approach is especially useful in an over-deployed network where one can find a minimal subset of the sensors that will maintain the desired level of coverage and power off all other nodes to reduce power consumption [47, 48, 52, 57].

5.2.1 Coverage Holes

The problem of locating coverage holes in a coordinate-free network has been studied in [13–15]. A strictly graph theoretic approach is used to locate nodes on the boundary of holes; however, no clear definition is given of what constitutes a hole for the proposed method. Additionally, [14] provides a graph theoretic method for using only local connectivity information (i.e., no coordinates) to develop a rough sketch of the layout of the network. The relationship between sensing coverage and node connectivity in sensor networks is given in [57]. This fundamental result states that *if the communication range of the sensor nodes is at least twice the sensing range then complete sensing coverage of a convex region implies connectivity of the network.* A generalization of this result to higher degrees of coverage, often referred to as *k*-coverage, is presented in [52]. A sensor field is called *k*-covered if every point is covered by at least *k* sensors. Under the same assumption, that the communication range be at least twice the sensing range, *k*-coverage implies *k*-connectivity.

Local connectivity information, without any positioning information, was used in [19, 40, 45] to determine if a region is covered or not [23]. Mathematical models of homology [45] were used to determine the coverage. A further extension of this work [19] shows how to detect the boundary of a hole in the network. This method only works for sensor networks that have a single hole and for networks having multiple holes it likely will not identify all holes. Furthermore, there is no guarantee that the algorithm will exactly locate even single holes. These references present a method of determining coverage without knowing the positions of the sensors. The major limitation is that there must be a certain ratio between the sensing range and the communication range of the sensor nodes or connectivity information alone is not enough to imply sensing coverage.

Whereas much study has been focused on how to improve coverage in sensor networks, research in [25] seeks to identify any holes in coverage that may be present in the existing sensor network deployment. Only after verifying that coverage is not sufficient would attempts be made to improve coverage or patch holes.

If the area of interest is not completely covered, it is called *coverage holes*, or simply *holes*. It is in these holes that an event may go undetected. Figure 5.10 gives an example of a sensor network with fully 1-covered sensor field and a sensor network with a coverage hole.

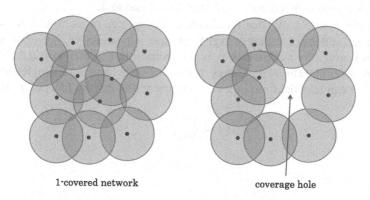

1-covered network   coverage hole

Fig. 5.10 Fully covered (1-covered) sensor field (*left*) and deployment with a coverage hole (*right*)

After determining the existence of holes it is also desirable to locate the holes so that, if necessary, they can be patched. Detecting the location of holes may be useful for other reasons as well. The presence of a hole may indicate a feature of the terrain, such as a lake, or some other phenomenon, such as a fire, which cannot be patched by the simple deployment of additional sensors. Also, messages are more likely to be routed through nodes on hole boundaries, depleting their available battery power more quickly, which could finally enlarge the hole [56] and degrade the network performance.

Assuming a simple, omnidirectional sensing model, each sensor node has a certain sensing range r_s and a transmission, or communication, range r_c.

In real-life applications, sensing and communication models are not always omnidirectional. Such models also heavily depend on environmental conditions.

Example 5.1 The IRIS sensor nodes are equipped with omnidirectional antennas capable of transmitting signal up to 500 m. That capability is produced in environments where there are no obstructions and the EM signal is allowed to propagate without interference. In the case of an urban environment, the sensing range would be considerably less; moreover, other factors such as multipath signal reflections, shadowing, and path loss due to non-free space propagation also affect the accuracy of the model. The received signal strength gradually decreases with the distance from the transmitter. The received signal strength is inversely proportional to the square of the distance between the transmitting and receiving antennas. Mathematically, it can be written as

$$\frac{P_r}{P_t} = \left[\frac{\sqrt{G_l}\lambda}{4\pi d} \right]^2, \tag{5.2}$$

where, P_r is the received power, P_t is the transmitted power, λ is the wavelength of the signal, d is the distance between receiver and transmitter, and G_l is the product of the transmit and receive antenna field radiation pattern. Because the antennas on the nodes are omnidirectional, the factor G_l is equal to 1.

Consider a set $S \subset \Re^2$ that represents a sensing area of interest. A network of N equal sensor nodes is deployed over S with sensor locations at (x_i, y_i, z_i). A sensing function is given by $p_i(q)$ where p_i is the probability that the event $q \in S$ will be detected. Such probability represents the sensing model and in its simplest form the probability can take the form of a uniform probability density function [20]

$$p_i(q) = \begin{cases} 1, & d_i \leq r_s \\ 0, & d_i > r_s \end{cases} \tag{5.3}$$

or

Fig. 5.11 Sensing and communication range of a sensor node

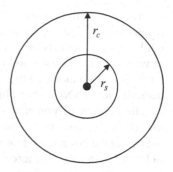

$$p_i(q) = \begin{cases} P_t \left[\frac{\sqrt{G_t}\lambda}{4\pi d_i} \right]^2, & d_i \leq r_s \\ 0, & d_i > r_s \end{cases} \tag{5.4}$$

where r_s is the sensing radius of the omnidirectional antenna and d_i is a distance between the location of a sensor node i and specific sensing point in the field q, i.e.,

$$d_i = \| (x_i, y_i, z_i) - q \|. \tag{5.5}$$

In case of a uniform sensing probability function, the radius r_s around each node is referred to as its sensing disk, see Fig. 5.11. It is within this disk that the sensor will detect the desired phenomenon with probability of one. This is sometimes referred to as a binary sensing model since an event is either detected or it is not detected. Probabilistic models are also discussed in the literature [21, 58] and may be more realistic than binary sensing model. The same is true for communication models that affect connectivity of WSNs. However, when coverage and connectivity are discussed, usually the simplest binary models are considered (sensors can either sense or not, nodes can either communicate with other nodes or not).

5.3 Connectivity

Deployment of wireless sensor networks requires minimization of cost, reduction in computation and communication, high-degree of sensing area coverage, while maintaining a connected network. If the network is not connected, information cannot flow from deployed sensors towards the network gateway or a base station. If a network is modeled as a graph, then network connectivity is equivalent to graph connectivity, i.e., there exists a path between any two vertices (nodes) in the graph. Equivalently to the degree of coverage, there is a degree of network connectivity where k-connectivity means that removing any $k - 1$ nodes still leaves the graph connected. Several deployment algorithms that optimize some aspect of coverage, while maintaining network connectivity, are given in [18]. Potential field algorithm provides sensor nodes mobility where coverage can be maximized, while the network is still connected [39]. Decentralized estimation and control of graph connectivity for WSNs with mobile nodes is given in [55].

The concept of potential field is an artificially created field where static and mobile nodes are subject to specifically designed forces. Since the goal is to maximize the coverage while maintain the connectivity, two forces were suggested in [39]: F^R—a repulsive force that move mobile sensor nodes away from each other to maximize the coverage; and F^A—an attractive force that pulls nodes towards each other to stay connected. Both forces are inversely proportional to the square of the distance between node pairs and are given by

$$F^R(i,j) = \frac{-K_r}{\left(\text{dist}\left[(x_i,y_i),(x_j,y_j)\right]\right)^2} \frac{(x_i,y_i) - (x_j,y_j)}{\text{dist}\left[(x_i,y_i),(x_j,y_j)\right]}$$

$$F^A(i,j) = \begin{cases} \frac{-K_a}{\left(\text{dist}\left[(x_i,y_i),(x_j,y_j)\right]-R_c\right)^2} \frac{(x_i,y_i)-(x_j,y_j)}{\text{dist}\left[(x_i,y_i),(x_j,y_j)\right]}, & \text{for critical connect.} \\ 0, & \text{otherwise} \end{cases} \tag{5.6}$$

where (x_i,y_i) is an x,y location of a sensor node i, and K_r and K_a are design constants. Nodes are initially positioned at one place, and then start repelling from each other. When they are only k-connected neighbors left for a single node (critical connection), the attractive force starts pulling sensor nodes, thus preventing further reduction in sensor nodes connectivity. The nodes reach equilibrium state when sum of all forces acting on a node is equal to zero. The method provides an elegant, distributed solution for deployment of mobile sensor nodes under k-connectivity constraints.

Distributed algorithm that applies virtual forces is presented in [22]. Such algorithm maximizes the coverage and maintains uniform distribution of sensor nodes. The method assumes that nodes have information about their coordinates. The virtual force is proportional to the distance between sensor nodes and *expected density* of sensor nodes in the sensing field. The force is given by

$$F(i,j) = \frac{\mu(i)}{\mu^2(R_c)} \left(R_c - \text{dist}\left[(x_i,y_i),(x_j,y_j)\right]\right) \frac{(x_i,y_i) - (x_j,y_j)}{\text{dist}\left[(x_i,y_i),(x_j,y_j)\right]}, \tag{5.7}$$

where $\mu(i)$ is the local node density the at sensor node i, and the expected density $\mu(R_c)$. The expected density depends on communication radius and is given by

$$\mu(R_c) = \frac{N\pi R_c^2}{A}, \tag{5.8}$$

with N being the number of sensor nodes, R_c communication radius, and A the sensing area. The algorithm provides movement of the nodes based on virtual forces and finishes when there are no new movements of nodes.

A centralized approach using virtual forces is given in [58, 60]. Each sensor has three virtual forces that act upon the node: repulsive force from the obstacles, attractive force from the sensing areas of high interest, and forces from other nodes. The specific forces from other nodes can either be repulsive or attractive, depending on their mutual distance and design parameters. The force applied to each sensor node is given by

$$F_i = \sum_{j \in N_i} F_{ij} + F_i^R + F_i^A, \qquad (5.9)$$

where F_{ij} is the force (repulsive or attractive) from node j to node i, F_i^R is the repulsive force applied to the node, and F_i^A is the attractive force applied to the node. The force F_{ij} is given by

$$F_{ij} = \begin{cases} w_A \left(\text{dist}\left[(x_i, y_i), (x_j, y_j)\right] - d_{th} \right) \angle \vec{ij}, & \text{if } \text{dist}\left[(x_i, y_i), (x_j, y_j)\right] > d_{th} \\ 0, & \text{if } \text{dist}\left[(x_i, y_i), (x_j, y_j)\right] = d_{th} \\ \frac{w_B}{\text{dist}\left[(x_i, y_i), (x_j, y_j)\right]} \angle \vec{ij} + \pi, & \text{if } \text{dist}\left[(x_i, y_i), (x_j, y_j)\right] < d_{th} \end{cases},$$

$$(5.10)$$

where $\vec{ij}$ is the vector from node i to node j, d_{th} is the threshold value for the mutual distance between nodes, and w_A and w_B are the constants determining amplitude of the forces. The method provides centralized deployment algorithm with controlled connectivity of the network. The sign of the force changes when distance between nodes crosses d_{th} threshold, allowing for controlled connectivity of the network. The computational complexity of the virtual force algorithm is $O(nmk)$ for a network of n nodes deployed on the $m \times k$ grid.

5.3.1   Graph Laplacian

The Laplacian matrix of a network communication graph is of fundamental importance in network connectivity, network diameter, mean distance, number of connected components of a graph and numerous other applications, [34]. We review a standard graph theory terminology that includes adjacency matrix and the Laplacian matrix of graphs. In Sect. 5.1.1 we defined an adjacency matrix $A(G)$ of a directional graph G as a matrix that at the position i, j has the number of edges from v_i to v_j. We also introduced a degree of a vertex v_i.

Let $D(G)$ be a diagonal matrix indexed by a vertex set $V(G)$, $D(G) = \text{diag}\{d(v_1), d(v_2), \ldots, d(v_n)\}$. The *Laplacian* matrix of a graph G is defined as

$$Q(G) = D(G) - A(G). \tag{5.11}$$

The Laplacian matrix is also called *Kirchhoff* matrix or admittance matrix and it finds its application in electrical networks. The Laplacian matrix is positive semi-definite and symmetric matrix with the additional property that all rows sum to zero (show this property as an exercise). The characteristic polynomial of $Q(G)$ is given by

$$\mu(G, \lambda) = \det(\lambda I - Q), \tag{5.12}$$

where I is the identity matrix of same dimensions at Q. The characteristic polynomial roots are the Laplacian eigenvalues $\lambda_1 \leq \lambda_2 \leq \cdots \leq \lambda_n$ where n is the order of G.

Example 5.2 Consider two graphs shown in Fig. 5.12.

The adjacency matrix A_1 (for the graph on the left) and matrix A_2 (for the graph on the right) are given by

$$A_1 = \begin{bmatrix} 0 & 1 & 1 & 1 & 0 \\ 1 & 0 & 0 & 1 & 0 \\ 1 & 0 & 0 & 1 & 1 \\ 1 & 1 & 1 & 0 & 1 \\ 0 & 0 & 1 & 1 & 0 \end{bmatrix}, A_2 = \begin{bmatrix} 0 & 0 & 1 & 0 & 0 \\ 0 & 0 & 0 & 1 & 0 \\ 1 & 0 & 0 & 0 & 1 \\ 0 & 1 & 0 & 0 & 0 \\ 0 & 0 & 1 & 0 & 0 \end{bmatrix}. \tag{5.13}$$

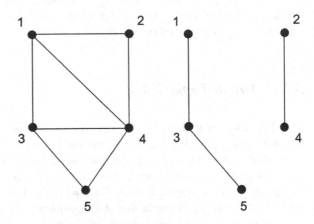

Fig. 5.12 Connected (*left*) and disconnected (*right*) graph

Corresponding Laplacian matrices are:

$$
Q_1 = \begin{bmatrix} 3 & 0 & 0 & 0 & 0 \\ 0 & 2 & 0 & 0 & 0 \\ 0 & 0 & 3 & 0 & 0 \\ 0 & 0 & 0 & 4 & 0 \\ 0 & 0 & 0 & 0 & 2 \end{bmatrix} - \begin{bmatrix} 0 & 1 & 1 & 1 & 0 \\ 1 & 0 & 0 & 1 & 0 \\ 1 & 0 & 0 & 1 & 1 \\ 1 & 1 & 1 & 0 & 1 \\ 0 & 0 & 1 & 1 & 0 \end{bmatrix}
$$

$$
= \begin{bmatrix} 3 & -1 & -1 & -1 & 0 \\ -1 & 2 & 0 & -1 & 0 \\ -1 & 0 & 3 & -1 & -1 \\ -1 & -1 & -1 & 4 & -1 \\ 0 & 0 & -1 & -1 & 2 \end{bmatrix} \tag{5.14}
$$

$$
Q_2 = \begin{bmatrix} 1 & 0 & 0 & 0 & 0 \\ 0 & 1 & 0 & 0 & 0 \\ 0 & 0 & 2 & 0 & 0 \\ 0 & 0 & 0 & 1 & 0 \\ 0 & 0 & 0 & 0 & 1 \end{bmatrix} - \begin{bmatrix} 0 & 0 & 1 & 0 & 0 \\ 0 & 0 & 0 & 1 & 0 \\ 1 & 0 & 0 & 0 & 1 \\ 0 & 1 & 0 & 0 & 0 \\ 0 & 0 & 1 & 0 & 0 \end{bmatrix}
$$

$$
= \begin{bmatrix} 1 & 0 & -1 & 0 & 0 \\ 0 & 1 & 0 & -1 & 0 \\ -1 & 0 & 2 & 0 & -1 \\ 0 & -1 & 0 & 1 & 0 \\ 0 & 0 & -1 & 0 & 1 \end{bmatrix} \tag{5.15}
$$

and corresponding Laplacian eigenvalues are:

$$
\mu(G_1) : 0, 1.58, 3, 4.41, 5 \tag{5.16}
$$

$$
\mu(G_2) : 0, 0, 1, 2, 3. \tag{5.17}
$$

Note that the graph G_2 is disconnected and has two components, and its first two eigenvalues are equal to zero. The following theorem [34] relates spectrum of the Laplacian graph matrix and graph connectivity.

Theorem 5.1 *Let G be a weighted graph with all weights nonnegative. Then:*

(a) *$Q(G)$ has only real eigenvalues,*
(b) *$Q(G)$ is positive semi-definite,*
(c) *Its smallest eigenvalue is $\lambda_1 = 0$ and a corresponding eigenvector is $[1, 1, \ldots, 1]^T$. The multiplicity of 0 as an eigenvalue of $Q(G)$ is equal to the number of components of graph G.*

From Theorem 5.1, it follows that $\lambda_1 = 0$, $\lambda_2 > 0$ if and only if the graph G is connected. Therefore, the eigenvalues of the graph Laplacian serves as a measure of graph connectivity. The number $\lambda_2(G)$ is also called *algebraic connectivity* of the graph G as it directly relates to its connectivity [12]. If $\lambda_2 = 0$ then the graph (network) is disconnected. If $\lambda_2 > 0$, the graph (network) is connected, with the second eigenvalue λ_2 being a measure of the graph (network) connectivity. The following Theorem [34] lists few useful bounds on Laplacian eigenvalues.

Theorem 5.2 *Let $G(V, E)$ be a graph of order n. The following bounds are valid*:

(a) $\lambda_2 \leq \frac{n}{n-1} \min\{d(v)\}$

(b) $\lambda_n \leq \max\{d(u) + d(v)\}$ where $uv \in E(G)$

(c) $\sum_{i=1}^{n} \lambda_i = 2|E(G)| = \sum_v d(v)$

(d) $\lambda_n \geq \frac{n}{n-1} \max\{d(v)\}$.

The following result describes behavior of eigenvalues when additional edges are inserted in the communication graph [34]. This corresponds to a scenario when nodes move closer to each other, thus increasing their connectivity.

Theorem 5.3 *The eigenvalues of graph G and $G' = G + e$ satisfy*

$$0 = \lambda_1(G) = \lambda_1(G') \leq \lambda_2(G) \leq \lambda_2(G') \leq \cdots \leq \lambda_n(G) \leq \lambda_n(G'). \qquad (5.18)$$

The number of spanning trees is also related to the spectrum of Laplacian. Let $\kappa(G)$ be the number of spanning trees of the graph G of order n. The number of spanning trees is related to eigenvalues by

$$\kappa(G) = \frac{1}{n} \lambda_2(G)\lambda_3(G)\ldots\lambda_n(G). \qquad (5.19)$$

The second eigenvalue of a graph Laplacian is closely related to the graph diameter (the longest shortest path between any two graph vertices, i.e., the largest number of vertices that must be traversed to travel from one vertex to another), which relates to the sensor network coverage area. The upper and lower bounds on graph diameter are given by [34]

$$\frac{4}{n\lambda_2(G)} \leq \text{diam}(G) \leq 2\left\lceil \frac{\Delta + \lambda_2(G)}{4\lambda_2(G)} \ln(n-1) \right\rceil, \qquad (5.20)$$

where Δ is the maximum degree in graph G, $\Delta = \max\{d(v)\}$.

Applications of Laplacian eigenvalue-based analysis include mobile sensor networks, robotics swarms, cooperative control, consensus networks, and others. Most applications utilize the fact that the measure of network connectivity is the second smallest eigenvalue of the Laplacian matrix of the graph. For instance, in [43], a decentralized connectivity maintenance control strategy for mobile robotic systems/networks is considered, i.e., a group of N agents with single-integrator dynamics

$$\dot{p}_i = u_i, \qquad (5.21)$$

where p_i and u_i are position and control input, respectively, for the i-th agent. In [53], the connectivity maintenance control algorithm is given by

$$u_i = \frac{\partial \lambda_2}{\partial p_i}. \qquad (5.22)$$

To improve the system stability from the connectivity point of view, it is proposed in [43] a modified control algorithm where control signal increases as the algebraic connectivity of the graph decreases (graph becomes "*less connected*"). The modified controls algorithms is given by

$$u_i = \operatorname{csch}^2(\lambda_2 - \varepsilon)\frac{\partial \lambda_2}{\partial p_i}, \qquad (5.23)$$

where csch is the standard hyperbolic cosecant function and ε is the desired lower-bound for the value of λ_2. This approach allows the control signal to increases when connectivity in the network is reduced.

5.4 Coverage Models Using Voronoi Diagrams

In Chap. 2 we provided an overview of Voronoi diagrams. Voronoi diagrams can be used for studying WSN deployment, coverage control, and hole patching. A vector-based algorithm was proposed in [51] where sensors move from densely covered areas to sparsely covered areas. Sensors act on each other with repulsive force. The virtual force from neighboring sensors will move the sensor such that the mutual distance is close to average distance between nodes. During each movement, each sensor calculates future coverage within its Voronoi cell. If the future movement will not improve the coverage, the sensor will not move to the target location and will instead move to the midpoint position between its target location and new location. Voronoi cell also has repulsive forces from the cell boundary, thus pushing the sensor node toward the inside of the region.

Voronoi-based algorithm moves sensor nodes towards the maximum coverage. It is a greedy algorithm where each sensor checks for holes within its Voronoi cell. If the hole is detected, the node moves towards its farthest vertex in the Voronoi cell. The distance between the farthest Voronoi vertex and its new location is equal to the sensing range with a maximum moving distance of the node being equal to the half of the communication range.

A combination of mobile and static nodes offers reduced cost while networks still has flexibility of mobile networks. Voronoi diagrams can be used for a hole detection in such networks [50]. Static nodes map their Voronoi diagrams and calculate coverage holes within their Voronoi cells. Static nodes then start the bidding process in which a mobile node will move to patch the hole. If there is a hole, a static sensor chooses the Voronoi vertex that is the farthest from the sensor and calculates related bid. The bid is given by

$$B_i = \pi(d - r_s)^2, \tag{5.24}$$

where d is the distance between the sensor node and the Voronoi vertex, and r_s is the sensing range. Mobile nodes also have their own base price. The static node finds the mobile node whose base price is lower than its bid. The mobile node receives all bids from its neighbors, and based on the highest bid moves to heal the coverage hole. Once the mobile node accepts the bid, it updates its base price with that specific bid. This bidding-price model guarantees that the total size of holes will decrease over time, meaning no mobile node will patch the one hole, and create larger one at the same time.

5.5   Simplicial Complexes

A k-simplex is defined as the convex hull of a set of $k + 1$ points in $\Re^n$. More intuitively, a k-simplex is simply an unordered $(k + 1)$-tuple of points. For example, a 0-simplex (σ_0) is defined as a single point and a 1-simplex (σ_1) is defined as an unordered pair of points. These are identical to vertices and edges in graph theoretic terminology. A 2-simplex (σ_2), comprised a triple of points is informally a triangle that is "filled in" so that it is not hollow inside. A 3-simplex (σ_3), also called a tetrahedron, is defined by a set of four points, and is again, solid inside. In general, any k-simplex has a solid interior, though it is difficult to visualize this in higher dimensions. Examples of the first several k-simplices are illustrated in Fig. 5.13 [40].

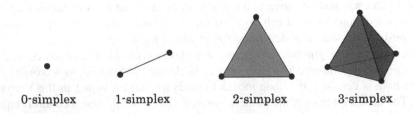

0-simplex   1-simplex   2-simplex   3-simplex

Fig. 5.13 Examples of a 0-simplex, a 1-simplex, a 2-simplex, and a 3-simplex

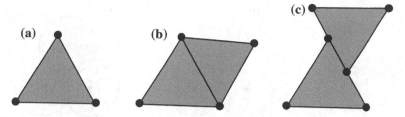

Fig. 5.14 Examples of simplicial complexes (**a, b**) and a non-simplicial complex (**c**)

This concept of k-simplices may also be extended to include collections of attached simplices, which are known as *simplicial complexes*. A simplicial complex K is a finite collection of simplices that satisfies the following two conditions:

1. For any simplex $\sigma_i \in K$, each face of σ_i is also in K;
2. Two k-simplices $\sigma_1, \sigma_2 \in K$ must intersect at a common face.

While (a)–(c) in Fig. 5.14 are collections of k-simplices; it is clear from the definition that (c) is not a simplicial complex because it violates condition 2.

It is evident from the above discussion that there is a strong similarity between graphs and simplicial complexes. In fact, a simplicial complex is in a sense just the generalization of a graph to higher dimensions. If we define the *r-skeleton* of the simplicial complex K as the collection of all k-simplices of K for which $k \leq r$, then it is clear that the 1-skeleton of K is a graph, also called the *underlying graph* of K, since it is simply a set of vertices and edges.

5.5.1 From WSNs to Simplicial Complexes

We introduce here two important simplicial complexes and their respective planar counterparts and show how each may be used to model a WSN [40]. The sensor network is mapped to one of these simplicial complexes which correspond to sensing coverage. Then, the simplicial complex can be analyzed to find the number and location of holes, which relate to holes in the actual sensor network. Such simplicial complexes can model the coverage of the sensor networks and related quality of service in terms of coverage completeness.

Čech Complex The Čech complex, also known as the nerve complex [4], can be defined for a set of points X with a parameter $r > 0$. The points $x_0, x_1, \ldots, x_n$ are defined as 0-simplices. The 1-simplex $[x_0 x_1]$ is defined if the r-balls centered at x_0 and x_1 intersect. If the r-balls centered at x_0, x_1, and x_2 have a common intersection, then the 2-simplex $[x_0 x_1 x_2]$ exists in the Čech complex. In general, the k-simplex $[x_0 x_1 x_2 \ldots x_k]$ exists if and only if the r-balls centered at x_0, x_1, $\ldots$, x_k have a nonempty common intersection.

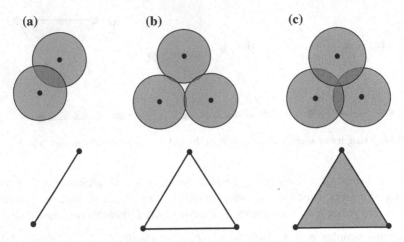

Fig. 5.15 Constructing the Čech complex from the union of sensing disks

It is obvious that this construction fits well for modeling coverage in a WSN. To observe this, one only needs to consider that r from the above construction equals the sensing radius r_s. For example, when three sensing disks have a common intersection we add a 2-simplex to the Čech complex which signifies that the entire area inside the polygon (i.e., triangle) formed by these three sensor nodes is covered. The inclusion of higher dimensional simplices indicates higher degrees of coverage. Figure 5.15 illustrates the construction of simplices from the union of coverage disks. In (a), a 1-simplex is constructed since the disks have a common intersection. The disks in (b) do not have a common intersection and the corresponding triangle is not filled, while the disks in (c) do have a common intersection and the corresponding filled triangle is constructed (2-simplex) [40].

An important benefit of this simplicial complex is that it can be computed even if the nodes have different sensing ranges, as long as the sensing model is a circle around the node. For theoretical analysis purposes it is usually assumed that all nodes have identical sensing ranges, but do note that this assumption is, in general, not required. However, a potential disadvantage of this simplicial complex construction is that precise distances between nodes must be known. If these distances are not precisely known this could lead to errors in the constructed Čech complex.

Rips Complex [40] The Rips complex, sometimes also called the Vietoris complex, was originally introduced by Vietoris in 1927 [49]. Given the point set X with a diameter r, the Rips complex is defined as follows. The k-simplex $[x_0 x_1 x_2 \ldots x_k]$ is defined in the Rips complex if the pairwise distance between each of x_0, x_1, x_2, ..., x_k is less than r. For example, when the distance between two points x_0 and x_1 is less than r, then we define the 1-simplex $[x_0 x_1]$. Likewise, if the pairwise distance between the three points at x_0, x_1, and x_2 are all less than r, then we define the 2-simplex $[x_0 x_1 x_2]$.

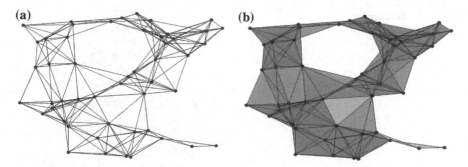

Fig. 5.16 Communication graph (**a**) and its induced Rips complex (**b**)

Note that the definition of the Rips complex is very similar to that of the unit disk graph from Sect. 5.1.1. It is apparent from the two definitions that using parameter r for each, the 1-skeleton of the Rips complex is equivalent to the unit disk graph. The Rips complex can be considered as a sort of "higher dimensional" version of the unit disk graph.

This similarity in the Rips complex and the unit disk graph leads to an alternate approach to constructing the Rips complex which is extremely useful for sensor networks and was exploited in [45] to construct the Rips complex in a coordinate-free network. Given the communication graph of the network, which by definition is a unit disk graph, the graph can be expanded to include higher dimensional simplices as follows. Any complete graph on k vertices, that is, the graph in which all k vertices are pairwise adjacent, corresponds to a $(k-1)$-simplex in the Rips complex. Thus, any triangles (K_3) in the communication graph are "filled in" to become 2-simplices, any tetrahedrons (K_4) are "filled in" to become 3-simplices, and so on. Since no locations are needed to create the communication graph the Rips complex for a sensor network can be constructed entirely from local connectivity data. Figure 5.16 shows an example communication graph and its induced Rips complex.

5.5.2 Comparison of Čech Complex and Rips Complex

While the Čech complex maps the coverage provided by the union of sensing coverage disks into a corresponding simplicial complex, this relationship is not as well defined for the Rips complex whose construction is based solely on the communication range of the nodes and does not take into account their sensing range at all. In [45], the ratio $r_c \leq \sqrt{3} \cdot r_s$ is used since for a set of three nodes with maximum pairwise distance for communication range of r_c, the sensing radius must be at least $r_c/\sqrt{3}$ for the sensing disks to have common intersection as shown in Fig. 5.17 [40].

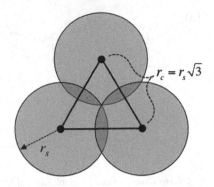

$$r_c = r_s\sqrt{3}$$

r_s

It can also be shown that the radius-r Rips complex has the same 1-skeleton as
the radius-$r/2$ Čech complex. Thus, assuming $r_s = r_c/2$, then the Rips complex and
the Čech complex are equivalent with the possible exception with three communicating nodes without a common intersection of their sensing disks. In other
words, triangle-shaped holes, such as that in Fig. 5.15b, are not possible in the Rips
complex since based on the above assumption the nodes are all within communication range, and they would immediately be "filled in" as 2-simplices.

Figure 5.18 shows examples of Rips and Čech complexes created using the
above assumption. Notice that the only difference is that the triangle-shaped holes
in the Čech complex do not exist in the Rips complex as mentioned above. This
could be acceptable in some applications since the triangular holes represent areas
with very small coverage holes, and larger holes will exist in both complexes. The
drawback to using this assumption, however, is that if this ratio is not exact, then
this equivalence between two simplicial complexes will not hold.

Even with the help of either of the above assumptions, the Čech complex and
Rips complex are still not equivalent. For this reason, results in [45] instead show
how to infer Čech data by "squeezing" it between two Rips complexes constructed
at different radii. This idea may be useful for verifying coverage, but it is not as

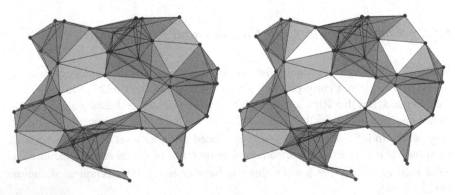

Fig. 5.18 Rips complex (*left*) and Čech complex (*right*) for $r_s = r_c/2$

helpful in localization of the holes. If the sensor nodes' coordinates are known, then the Čech complex may be used to capture the exact coverage of the sensing disks. Otherwise, if sensor nodes' coordinates are not known, then for any ratio between sensing and communication radii the Rips complex will only approximate the Čech complex.

5.5.3 Subcomplexes with Planar Topology

We introduce subcomplexes of both the Čech complex and the Rips complex [40]. The significance of these subcomplexes is that their underlying graphs are planar, meaning that they have no crossing edges. As a result, both simplicial complexes are 2-complexes, which mean that the largest k-simplex in the complex is a 2-simplex and k-simplices of higher dimensions cannot exist in the complex. These subcomplexes can be considered as the maximal subcomplexes that retain the number and shape of the holes in the original complex.

Alpha-Shape Complex [40] Closely related to the Čech complex is the Alpha-shape complex, which is originally due to Herbert Edelsbrunner [11]. To construct this complex, one must first compute the Voronoi diagraph Vor_X for the point set X. Next, for each point $x \in X$ its Alpha-cell $A(x, r)$ can be defined as the intersection $\mathrm{Vor}_X(x) \cap B(x, r)$, where $\mathrm{Vor}_X(x)$ is the Voronoi cell of x and $B(x, r)$ is an open ball of radius r centered at x. The Alpha-shape complex $C_\alpha(X, r)$ is defined by the nerve, or Čech complex of the set of Alpha-cells $A(x, r)$. That is, a k-simplex in the Alpha-shape complex corresponds to the non-empty intersection of $k + 1$ alpha-cells.

Figure 5.19 shows the Voronoi diagram and sensing disks for a set of points. The collection of Alpha-cells, which is the intersection of each sensing disk and its Voronoi cell, is shown in Fig. 5.20. Finally, the nerve of these alpha-cells is taken to create the alpha-shape complex shown in Fig. 5.20. Notice that the two holes in the alpha-shape complex clearly correspond to the holes in the union of sensing disks from Fig. 5.19.

It can be shown that for a given parameter r and a set of points X there is a homotopy equivalence between the two complexes, i.e., the thickness can be ignored (i.e., higher dimensional k-simplices), and the holes stayed preserved in the complex as shown in Fig. 5.21. Also, as may be expected with the use of the Voronoi diagram, the Alpha-shape complex is a subcomplex of the Delaunay complex and so its underlying graph must be planar.

Maximal Simplicial Complex [40] Note that it is also highly desirable to have a planar subcomplex of the Rips complex, analogous to the relationship between the Čech and Alpha-shape complexes. The basic goal is to remove a subset of the edges in the communication graph such that the new graph is planar and the number and

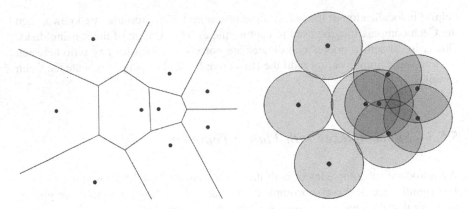

Fig. 5.19 Voronoi diagram (*left*) and union of sensing disks (*right*) for a set of points

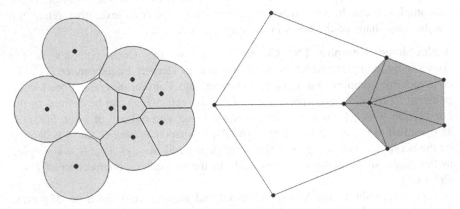

Fig. 5.20 Alpha cells (*left*) and alpha shape complex (*right*) for point set from Fig. 5.19

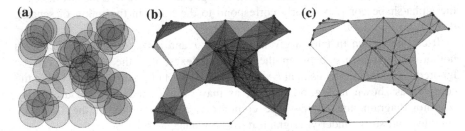

Fig. 5.21 Union of sensing disks (**a**), Čech complex (**b**), and alpha shape complex (**c**) for a set of sensor nodes

shape of holes is maintained in the induced Rips complex. In general, this is a difficult problem since the embedding is not known and removal of the wrong edges will have an effect on the holes in the induced Rips complex.

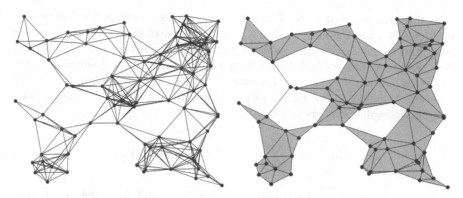

Fig. 5.22 Communication graph (*left*) and its maximal simplicial complex (*right*)

Researchers in [35] have proposed a solution to this problem by examining all crossing edges in the communication graph and carefully eliminating one edge in the crossing pair while still maintaining the structure of the topological holes. Authors refer to the planar subgraph as a *maximal simplicial subgraph*. After obtaining this subgraph, all that is left is to "fill in" all the triangles to obtain the induced Rips complex, which they call the *maximal simplicial complex*.

Figure 5.22 shows an example communication graph and its maximal simplicial complex obtained from the communication graph. The relative position of holes is maintained in the maximal simplicial complex.

5.6 Simplicial Homology and Coverage Holes

Homology is a mathematical method for detecting and counting holes in a topological space using algebra. A basic background on homology theory that is relevant to the sensor networks coverage is presented here, while a more complete development of homology theory can be found in [24, 36].

One way to define homology is as a vector space. For a simplicial complex K, the 1-dimensional homology $H_1(K)$ can be used to determine its holes. This can be represented as a vector space whose basis represents cycles in K surrounding its holes. Thus, the dimension of this vector space $|H_1(K)|$, often called the first Betti number of K, gives the number of cycles that corresponds to the number of holes. Holes of higher dimension may also be found using homology. For example, the second Betti number for a sphere S, which is the dimension of $H_2(S)$, is 1 because the sphere is empty inside. For the sensor network coverage problems, one-dimensional holes correspond to holes in sensing coverage.

Computing the first Betti number of a simplicial complex is important in the analysis of coverage completeness since it provides information about a number of holes that exist in the network. Two software packages for computional homology are PLEX [46] and CHomP [8]. The typical Betti number implementations involve computing the rank of matrices, which for very large networks, can be prohibitively slow. However, this method does not take into account that the simplicial complexes are not arbitrary but are constrained by the physical nature of the sensor network. A simple arithmetic formula for computing the first Betti number of a simplicial complex K is given by:

$$b_1 = b_0 - \sum_{k=0}^{n} (-1)^k |(k\text{-simplices in } K)|, \tag{5.25}$$

where n corresponds to the largest k-simplex in K, and b_0 is the zeroth Betti number, which is equivalent to the number of connected components in the underlying graph, or 1-skeleton, of K. This follows from the Euler characteristic χ for a simplicial complex which can be defined as the alternating sum of Betti numbers

$$\chi = \sum_{k=0}^{\infty} (-1)^k b_k. \tag{5.26}$$

For an arbitrary simplicial complex, Betti numbers of any dimension may be nonzero. However, all the simplicial complexes that relate to sensor networks have constraints due to the disk-shaped sensing and communication ranges which have been assumed. Therefore, all Betti numbers of dimension larger than one will be zero if the nodes lie in the plane. This leads to the simplified formula

$$\chi = b_0 - b_1, \tag{5.27}$$

where b_0 is equivalent to the number of connected components in the underlying graph. A second definition of the Euler characteristic for a simplicial complex K using the alternating sum of the number of k-simplices is given by

$$\chi = \sum_{k=0}^{n} (-1)^k |(k\text{-simplices in } K)|, \tag{5.28}$$

where n corresponds to the largest k-simplex in K. From here follows the Eq. (5.25).

The drawback of (5.25) in computing coverage holes is that one must know all higher dimensional simplices, whereas with the matrix intensive computations, only the $(k + 1)$-simplices and lower dimensions are needed to find the k-th Betti number. Thus, only the sets of 0-, 1-, and 2-simplices would be needed to compute b_1.

However, for the planar subcomplexes described above, the following simplified result can be used:

$$b_1 = b_0 - |V(G)| + |E(G)| - |(2\text{-simplices in } K)|, \qquad (5.29)$$

where G is the underlying graph of the simplicial complex K. Under assumption that the 1-skeleton of K is a connected graph, which is typically the case, then $b_0 = 1$ and so the computation can be simplified even further to

$$b_1 = 1 - |V(G)| + |E(G)| - |(2\text{-simplices in } K)|. \qquad (5.30)$$

Note that this formula can also be derived from Euler's formula for connected planar graphs. Euler's formula for a connected planar graph can be written as $V - E + F = 2$, where V is the number of vertices, E is the number of edges, and F is the number of faces. This formula, however, accounts for an additional face known as the infinite, or exterior face, which is not needed for sensor networks applications and can be discarded. Also, it is clear that the number of faces F in the planar underlying graph of one of the planar subcomplexes is the sum of the number of 2-simplices in the complex and the number of holes. This is due to the obvious fact that every face in the underlying graph is either a 2-simplex or a hole.

Example 5.3 Consider the simple complex illustrated in Fig. 5.23. The simplicial complex has seven 0-simplices (or vertices), nine 1-simplices (or edges), and two 2-simplices. Inserting these values into Eq. (5.30), we see that $b_1 = 1 - 7 + 9 - 2 = 1$, as expected by observation of the simplicial complex (Fig. 5.23).

In applications related to sensor networks hole coverage, the alpha-shape complex and the maximal simplicial complex have advantage over other complexes for multiple reasons. First, algorithms for computing the number of holes in the complex are computationally efficient by utilizing simple arithmetic calculations. Second, divide-and-conquer algorithms are often suitable for planar graphs. Furthermore, it allows for coverage holes to be defined as faces in the underlying planar graph. If the underlying graph is not planar, holes may not be uniquely identifiable as is shown in Fig. 5.24. Notice that the hole is not uniquely defined since one could argue that cycle A, B, D, cycle A, B, C, or cycle A, B, C, D are all the correct way to identify the hole.

Fig. 5.23 Example simplicial complex with $b_1 = 1$

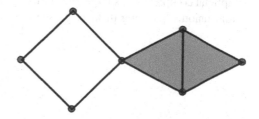

Fig. 5.24 Sensing disks (*left*)
and corresponding simplicial
complex (*right*) which
contains a hole that is not
uniquely identifiable

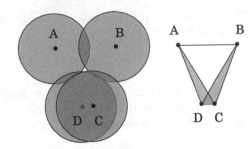

5.7   *K*-Coverage

Coverage in WSNs applications is usually associated with 1-coverage where it is
desired to have *at least one* sensor node covering every point of interest. If
1-coverage is not satisfied, then there is a hole in the sensor network coverage. The
term *hole* here also refers to a 1-hole, meaning there is not at least one sensor node
covering a certain area in the set of interest. A more robust network to node failures
and other node related errors might have *k*-coverage requirements, i.e., that every
point in the area of interest should be covered by at least *k* sensor nodes.

Definition 5.7 Consider a sensor network consisting of a set Z sensor nodes and a
region on interest S. A subset $P \subset S$ is said to be *k*-covered if and only if for every
point $p \in P$, there are at least k sensor nodes from Z that cover the point p.

Looking at the construction of the Čech complex it is evident that only a single
point must be covered by $k + 1$ disks to create a *k*-simplex (e.g., see Fig. 5.15c). It
is required to show that the entire area inside a polygon formed by k vertices is
k-covered. This implies that for any single polygon vertex the other $k - 1$ vertices
must all lie within its sensing disk and leads us back to the Rips complex, which
admits a *k*-simplex if the pairwise distance between any set of k vertices is less than
parameter r. Thus, a construction of the Rips complex using the parameter r_s allows
one to determine *k*-coverage. For example, any 3-simplex in this simplicial complex
comprised four vertices and, of course, the area inside the polygon formed by these
vertices is 4-covered. Note that this method gives a sufficient (but not necessary)
condition for *k*-coverage since there may still be other *k*-covered portions of the
network which do not entirely fill the area inside a polygon formed by a subset of
nodes. However, this is much like the case for 1-coverage where a hole in the
simplicial complex looks large, but in reality, is much smaller when considering the
actual union of sensing disks (e.g., see Fig. 5.15b).

5.8 Coverage Control

The problem of deploying sensors to satisfy application requirements is called coverage control problem [5]. The coverage control can be static, or offline, and dynamic, or online. In the static coverage, the sensor node locations are pre-calculated before the deployment, such that once deployed, the sensor network satisfies pre-assigned coverage mission requirements. This is equivalent to the facility location optimization problem in operations research. The dynamic coverage control includes mobile sensor networks, Fig. 5.25, where the coverage control is calculated online and the position of sensor nodes is adjusted according to specified criteria. Furthermore, the dynamic coverage control can be centralized with a central controller (or a base station) providing control inputs to the whole network, or distributed where each sensor node adjusts its position according to the locally executed algorithm.

Another example of dynamic coverage control is a future WSN usage in weather monitoring and hurricane tracking. Even though hurricanes can be tracked by satellites, precise wind speed, precipitation, and pressure of the storm-affected area can be monitored and detected only using close-range sensing devices. Presently such monitoring and data gathering is done using NOAA reconnaissance airplane with probes that record air pressure, humidity, temperature, and wind speed that weather scientists use to predict storm surge, place and time when the storm hits the land.

Future hurricane monitoring systems will consist of a network of ground sensors and mobile sensors with coverage that can dynamically adapt to changes in the hurricane path and strength; see Fig. 5.26.

Several optimal control problems related to sensor networks were formulated in [5]. An optimal coverage under constraints of imprecise detection and terrain properties where the number of sensor nodes is minimized was presented in [9].

Fig. 5.25 Micro-aerial vehicle based sensor network—a mobile sensor networks with dynamic coverage control

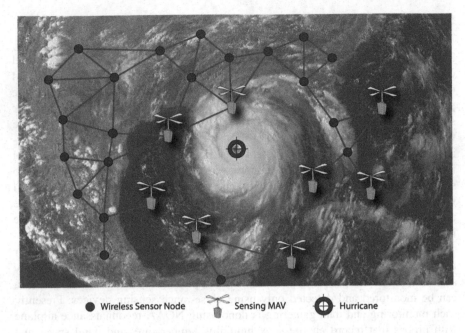

Fig. 5.26 Static and mobile sensor nodes track a hurricane

In [17] problems arising in maintaining coordination and communication between the group of robots and solutions to these problems were discussed. Three models of deployment are introduced to maximize the coverage area within the close range of the mobile nodes, deployment to maximize the probability of detecting a source, and deployment to maximize the visibility of the network. A variety of control methods in multi-vehicle cooperative control using graph theory have been presented in [41]. Optimal coverage control for mobile sensor networks was presented in [10]. The paper uses a Voronoi partitioning and Lloyd descent algorithm but does not consider network connectivity constraints. Two location functions that characterize coverage performance were provided in [9] including a study of their gradient properties via nonsmooth analysis. In most cases, the feasibility sets are assumed to be convex and related optimization problems are convex optimization problems. If network connectivity is considered as well, then underlying optimization problems become, in most cases, non-convex optimization problems.

Example 5.4 Consider a coverage control problem in chemical plants where it is required to provide an optimal coverage of large chemical plants or areas of interest with a large number of static sensor nodes that monitor for a wide spectrum of chemical agents; have several mobile nodes moving over rough plant terrain and track a possible contamination cloud; and allow technicians to adjust the controller of all mobile sensor nodes. This is an optimal coverage control problem with a trade-off between uniform coverage of the whole plant and focused coverage of the contamination cloud; see Fig. 5.27.

Let r be the sensors radio transmission range. Assume that there is a focus point $(X_F(t), Y_F(t))$ where several mobile sensor nodes should converge. In the case of a chemical contamination example, the focus point can be a contamination cloud (center of mass of the cloud). Consider a region of interest S that is a compact set, and a set Z of sensor nodes. A subset $M \subset Z$ is a set of all mobile sensor nodes. An optimal coverage control problem can be formulated by specifying a cost function of interest and constraints that limit the network in terms of geometry, flow, energy, or any other network parameter.

The coverage control problem can be formulated as an optimal control problem, or more precisely, as a linear-quadratic regulator (LQR) problem. The problem is to find an optimal location of mobile nodes M, such that the following cost function is minimized:

$$\min J_1 = R \sum_{i \in M} \text{dist}^2\left[(x_i, y_i), (X_F(t), Y_F(t))\right] + Q \sum_{i \in M, j \in Z} \text{dist}^{-2}\left[(x_i, y_i), (x_j, y_j)\right],$$

$$(5.31)$$

where R and Q are the control design parameters or weighting factors, and $\text{dist}\left[(x_i, y_i), (x_j, y_j)\right]$ is an Euclidean distance between nodes i and j and is given by

$$\text{dist}\left[(x_i, y_i), (x_j, y_j)\right] = \sqrt{(x_i - x_j)^2 + (y_i - y_j)^2}. \qquad (5.32)$$

Note that this is a convex optimization problem since the optimization cost function is convex and there are no constraints. Convex optimization refers to cost function being a convex function and constraint set being a convex set; otherwise, optimization is a non-convex and common operation research methods do not apply. For extra reading on non-convex optimization, we refer to [27]. The parameter R penalizes closeness to the focus point, or tracking of the target. The parameter Q penalizes uniform distribution of nodes across the set S. For example,

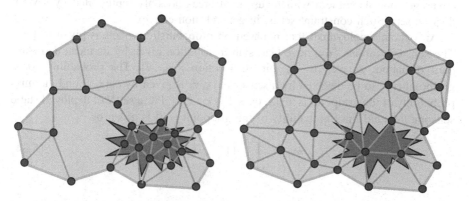

Fig. 5.27 Focused coverage (*left*) versus uniform coverage (*right*)

$R = 10$, $Q = 10,000$ means that the user is much more concerned with the uniform node distribution than to cover the specific area of interest, Fig. 5.27 (right). On the other hand, $R = 10,000$, $Q = 10$ indicates that the user wants extensive coverage of the focus point (tracking) rather than uniform node distribution, Fig. 5.27 (left).

The cost function in (5.31) has such specific form because the original problem is a multi-objective optimization problem. One common way to solve this type of problem is to combine the weighted sums of each individual objective function into an aggregate objective function (AOF). Specifically, the two objectives are to minimize the distance between the mobile nodes and the focus point and to maximize the distance between the mobile nodes and the other static and mobile nodes (minimize the inverse of this distance).

The problem formulation, including the cost function that is given in (5.31) has more of a theoretical significance, since it does not consider the network connectivity. Namely, such a solution will optimize the cost function and distribute the nodes accordingly without consideration of the network connectivity. Some sensor nodes can go out of communication range and become useless from the network standpoint. More realistic problem formulation that includes network connectivity can be given as follows.

Given the compact coverage area of interest S, a focus point $(X_F(t), Y_F(t))$, sensor network graph $G(t) = (N(t), E(t))$ and mobile subgraph $G_m(t) = (M(t), E_m(t))$, find an optimal vertex location of the mobile subgraph $G_m(t)$ such that the following cost function is minimized

$$\min J_2 = R \sum_{i \in M} \text{dist}^2 [(x_i, y_i), (X_F(t), Y_F(t))] + Q \sum_{i \in M, j \in N} \text{dist}^{-2} [(x_i, y_i), (x_j, y_j)],$$

(5.33)

such that the graph $G(t)$ stays vertex k-connected and satisfies the mobile node localization condition. Constants R and Q are control design parameters.

Note that this is a non-convex optimization problem where a feasibility set is, in general, a non-convex set. Similarly, one can formulate suboptimal sensor network coverage control problem with focus/target areas possibly represented by several disjoint sets. Such constraint set is, in general, non-convex.

An optimal deployment that is based on probabilistic models is given in [59]. The sensor locations (x, y) have Gaussian distribution given by its mean and standard deviation, as well as its join distribution $p_{xy}(x, y)$. The probability for a gridpoint (i, j) to be detected by a sensor at (x, y) is given by $c_{ij}(x, y)$ and the miss probability of the gridpoint is $m_{ij}(x, y) = 1 - c_{ij}(x, y)$. For a set N of deployed static sensors, the total miss probability of the gridpoint (i, j) is given by

$$m_{ij} = \prod_{(x,y) \in N} (1 - c_{ij}(x, y)).$$

(5.34)

Assuming that the newly deployed sensor will be placed at (x, y), the total miss probability is given by

$$m(x, y) = \sum_{(i,j)\in\text{Grid}} m_{ij}(x, y)m_{ij}. \tag{5.35}$$

Based on the calculated total miss probability, one can place the sensor to minimize the miss probability, providing the best possible sensor network coverage from the probability of detection point of view under given assumptions of grid deployment. After every node deployment, above probabilities are updated and recalculated based on new topology. The computational complexity of miss probability algorithm is $O(n^2m^2)$ for a grid size of $n \times m$.

Questions and Exercises

1. What is the degree of a vertex in a graph G? Provide an example.
2. Explain when is a graph connected? Draw examples of connected and disconnected graphs.
3. Demonstrate by example that if a graph G is disconnected, then its complement $\bar{G}$ is connected. Show another example that the opposite is not true—if the graph G is connected, then its complement $\bar{G}$ is not necessary disconnected. Finally, prove this statement rigorously.
4. Study the Chvátal's art gallery theorem and find out what is an upper bound on the minimal number of guards that are required to cover the gallery? Please describe "upper bound on the minimal number" in case of this problem.
5. If the gallery floor plan can be divided into square rooms, what would be the upper bound on the minimal number of guards required to cover the gallery in this case?
6. When one can say that complete sensing coverage of a convex region implies connectivity of the network?
7. Given a graph $V = \{1, 2, 3, 4, 5, 6, 7\}$, $E = \{(1, 2), (1, 3), (1, 5), (2, 3), (3, 5), (4, 7), (6, 7)\}$, using a graph Laplacian check if the graph is connected. Find the number of spanning trees using Laplacian eigenvalues.
8. Given a graph $V = \{1, 2, 3, 4, 5, 6, 7, 8, 9\}$, $E = \{(1, 2), (1, 3), (1, 5), (2, 3), (3, 5), (3, 6), (4, 5), (5, 8), (5, 9), (6, 7), (6, 8)\}$, estimate the lower and upper bounds on the graph diameter.
9. For a graph with vertices $V = \{1, 2, 3, 4, 5\}$, draw a complex and a simplicial complex.
10. For a set of points in two-dimensional space with the following coordinates $X = \{(-1, -2), (1, 0), (2, 3), (3, -4), (-3, -5)\}$, draw a Voronoi diagram and derive alpha shape complex.

11. Describe a k-coverage in sensor networks and its relationship with the Rips complex. What does it mean by 2-coverage specifically and in terms of the Rips complex?
12. Use linear quadratic control problem setup similar to the one given in Eq. (5.31) and formulate a sensor network coverage control problem where it is required for mobile nodes to be attracted to the target node, to be repelled from the static sensor nodes, and to be repelled from each other (mobile nodes).
13. How can you extend the previous problem formulation and include constraints that would ensure collision avoidance between mobile sensor nodes?

References

1. S. Adlakha and M. Srivastava, "Critical density thresholds for coverage in wireless sensor networks," *IEEE Wireless Communications and Networking*, vol. 3, pp. 1615–1620, March 2003.
2. A. Agah, S.K. Das, K. Basu, and M. Asadi, "Intrusion detection in sensor networks: a non-competitive game approach," *Third IEEE International Symposium on Network Computing and Applications*, pp. 343–346, 2004.
3. A. Arora, R. Ramnath, E. Ertin, P. Sinha, S. Bapat, V. Naik, V. Kulathumani, H. Zhang, H. Cao, M. Sridharan, S. Kumar, N. Seddon, C. Anderson, T. Hermon, N. Trivedi, C. Zhang, M. Nesterenko, R. Shah, S. Kulkarni, M. Aramugam, L. Wang, M. Gouda, Y. Choi, D. Culler, P. Dutta, C. Sharp, G. Tolle, M. Grimmer, B. Ferriera, and K. Parker, "*ExScal*: elements of an extreme scale wireless sensor network," *Proc. Of the 11th IEEE International Conference on Embedded and Real-Time Computing Systems and Applications*, pp. 102–108, August 2005.
4. G. Carlsson and V. de Silva, "Topological approximation using small simplicial complexes," (Preprint), 2003.
5. C. G. Cassandras and W. Li, "Sensor networks and cooperative control," *European Journal of Control*, vol. 11, pp. 436–463, 2005.
6. S. Chaudhry, V. Hung, and R. Guha, "Optimal placement of wireless sensor nodes with fault tolerance and minimal energy consumption," *IEEE International Conference on Mobile Ad-hoc and Sensor Systems*, pp. 610–615, October 2006.
7. V. Chvátal, "A combinatorial theorem on plane geometry," *Journal of Combinatorial Theory, Series B*, vol. 18, pp. 39–41, 1975.
8. "CHomP: computational homology project," http://chomp.rutgers.edu.
9. J. Cortes and F. Bullo, "Coordination and geometric optimization via distributed dynamical systems," *SIAM Journal on Control and Optimization*, vol. 44, no. 5, 2005, pp. 1543–1574.
10. J. Cortes, S. Martinez, T. Karatas, and F. Bullo, "Coverage control for mobile sensing networks," *IEEE Transactions on Robotics and Automation*, vol. 20, no. 2, 2004, pp. 243–255.
11. H. Edelsbrunner, "The union of balls and its dual shape," *Discrete & Computational Geometry*, vol. 13, pp. 415–440, 1995.
12. M. Fiedler, Algebraic connectivity of graphs, Czech. Math. J., vol. 23, no. 98, 1973, pp. 298–305.
13. S. Funke and C. Klein, "Hole detection or: how much geometry hides in connectivity?," *SCG '06: Proceedings of the Twenty-Second Annual Symposium on Computational Geometry*, pp. 377–385, New York, NY, 2006.

14. S. Funke and N. Milosavljevic, "Network sketching or: how much geometry hides in connectivity? – Part II," *Proc. of the 18th ACM-SIAM Symposium on Discrete Algorithms (SODA)*, 2007.

15. S. Funke, "Topological hole detection in wireless sensor networks and its applications," *DIALM-POMC: Joint Workshop on Foundations of Mobile Computing*, 2005.

16. D.W. Gage, "Command control for many-robot systems," *Proc. of Nineteenth Annual AUVS Technical Symposium*, pp. 22–24, 1992.

17. A. Ganguli, S. Susca, S. Martinez, F. Bullo, and J. Cortes, "On collective motion in sensor networks: sample problems and distributed algorithms," *Proc. of the 44th IEEE Conference on Decision and Control, and the European Control Conference*, 2005, pp. 4239–4244.

18. A. Ghosh and S. Das, "Coverage and connectivity issues in wireless sensor networks," in *Mobile, Wireless, and Sensor Networks: Technology, Applications, and Future Directions*, Ed. R. Shorey, A.L. Ananda, M.C. Chan, and W.T. Ooi, John Wiley & Sons, Inc., 2006.

19. R. Ghrist and A. Muhammad, "Coverage and hole-detection in sensor networks via homology," *Proc. ISPN*, 2005.

20. A. Gusrialdi, T. Hatanaka, and M. Fujita, "Coverage control for mobile networks with limited-range anisotropic sensors," *Proc. 47th IEEE Conference on Decision and Control*, Cancun, Mexico, Dec. 2009.

21. M. Hefeeda and H. Ahmadi, "A probabilistic coverage protocol for wireless sensor networks," *Proc. of IEEE International Conference on Network Protocols (ICNP '07)*, pp. 41–50, Beijing, China, October 2007.

22. N. Heo and P.K. Varshney, "A distributed self-spreading algorithm for mobile wireless sensor networks," *Proc. IEEE Wireless Communications and Networking Conference, (WCNC'03)*, New Orleans, LA, March 2003, pp. 1597–1602.

23. A. Howard, M.J. Matraric, and G.S. Sukhatme, "Mobile sensor network deployment using potential fields: a distributed, scalable solution to the area coverage problem," *International Symposium on Distributed Autonomous Robotics Systems*, June 2002.

24. T. Kaczynski, K. Mischaikow, and M. Mrozek, *Computational Homology, Applied Mathematical Sciences 157*, Springer-Verlag, 2004.

25. J. Kanno, J. G. Buchart III, R. R. Selmic and V. Phoha, "Detecting coverage holes in wireless sensor networks," *Proc. 17th Mediterranean Conference on Control and Automation*, Thessaloniki, Greece, June 24–26, 2009.

26. K. Kar and S. Banerjee, "Node placement for connected coverage in sensor networks", *Proc. of WiOpt 2003: Modeling and Optimization in Mobile, Ad Hoc, and Wireless Networks*, 2003.

27. D. Klatte and B. Kummer, *Nonsmooth Equations in Optimization*, Kluwer Academic Publishers, Dordrecht, The Netherlands, 2002.

28. A.A. Kooshesh and B.M.E. Moret, "Three-coloring the vertices of a triangulated simple polygon," *Pattern Recognition*, vol. 25, pp. 443–444, 1992.

29. S. Kumar, T. H. Lai, and A. Arora, "Barrier coverage with wireless sensor networks," *Proc. of the 11th Annual International Conference on Mobile Computing and Networking*, Cologne, Germany, pp. 284–298, 2005.

30. A.K. Lédeczi, A. Nádas, P. Völgysi, G. Balogh, B. Kusy, J. Sallai, G. Pap, S. Dóra, and K. Molnár, "Countersniper systems for urban warfare," *ACM Transactions on Sensor Networks*, vol. 1, no. 2, pp. 153–177, Nov. 2005.

31. D.T. Lee and A.K. Lin, "Computational complexity of art gallery problems," *IEEE Transactions on Information Theory*, vol. 32, no. 2, pp. 276–282, March 1986.

32. M. Marengoni, B.A. Draper, A. Hanson, and R.A. Sitaraman, "System to place observers on a polyhedral terrain in polynomial time," *Image and Vision Computing*, vol. 18, pp. 773–780, 1996.

33. S. Meguerdichian, F. Koushanfar, M. Potkonjak, and M. Srivastava, "Coverage poblems in wireless ad-hoc sensor networks," *IEEE Infocom 2001*, pp. 1380–1387, April 2001.

34. B. Mohar, "The Laplacian spectrum of graphs," in *Graph Theory, Combinatorics, and Applications*, vol. 2, Ed. Y. Alavi, G. Chartrand, O. R. Oellermann, A. J. Schwenk, Wiley, 1991, pp. 871–898.

35. A. Muhammad and M. Egerstedt, "Connectivity graphs as models of local interactions," *Proc. of the 43rd IEEE Conference on Decision and Control*, pp. 124–129, December 2004.
36. J.R. Munkres, *Elements of Algebraic Topology*, Addison Wesley, 1993.
37. J. O'Rourke, *Art Gallery Theorems and Algorithms*, Oxford University Press, 1987.
38. R. Pandit and P.M. Ferreira, "Determination of minimum number of sensors and their locations for an automated facility: an algorithmic approach," *European Journal of Operational Research*, vol. 63, no. 2, pp. 231–239, 1992.
39. S. Poduri and G.S. Sukhatme, "Constrained coverage in mobile sensor networks," *Proc. IEEE Int. Conf. Robotics and Automation (ICRA '04)*, New Orleans, LA, April 2004, pp. 40–50.
40. S. Ramazani, J. Kanno, R. Selmic, and M. Brust, "Topological and combinatorial coverage hole detection in coordinate-free wireless sensor networks," *International Journal of Sensor Networks*, vol. 21, no. 1, pp. 40–52, 2016.
41. W. Ren, R. W. Beard, and E. M. Atkins, "Information consensus in multivehicle cooperative control," *IEEE Control System Magazine*, vol. 27, no. 2, April 2007.
42. R. Roman, J. Zhou, and J. Lopez, "Applying intrusion detection systems to wireless sensor networks," *Consumer Communications and Networking Conference*, vol. 1, pp. 640–644, January 2006.
43. L. Sabattini, N. Chopra, and C. Secchi, "On decentralized connectivity maintenance for mobile robotic systems," *50th IEEE Conference on Decision and Control*, Orlando, FL, USA, December 2011.
44. R. Selmic, J. Kanno, J. Buchart, N. Richardson, "Quadratic optimal control of sensor network deployment," *Cyberspace Research Workshop*, Shreveport, LA, November 2007.
45. V.de Silva and R. Ghrist, "Homological sensor networks," *Notices of the American Mathematical Society*, vol. 54, no. 1, pp. 10–17, Jan. 2007.
46. "Plex: simplicial complexes in MATLAB," website: http://comptop.stanford.edu/programs.
47. S. Slijepcevic and M. Potkonjak, "Power efficient organization of wireless sensor networks," *IEEE International Conference on Communications*, vol. 2, pp. 472–476, June 2001.
48. D. Tian and N. Georganas, "A coverage-preserving node scheduling scheme for large wireless sensor networks," *ACM Workshop of Wireless Sensor Networks and Applications*, Atlanta, GA, October 2002.
49. L. Vietoris, "Über den höheren zusammenhang kompakter Raume und eine klasse von zusammenhangstreuen abbildungen," *Mathematische Annalen*, vol. 97, pp. 454–472, 1927.
50. G. Want, G. Cao, and T. LaPorta, "A bidding protocol for deploying mobile sensors," *Proc. 11th IEEE Int. Conf. Network Protocols (ICNP'03)*, Atlanta, GA, Nov. 2003, pp. 80–91.
51. G. Wang, G. Cao, and T. LaPorta, "Movement-assisted sensor deployment," *Proc. IEEE InfoCom*, Hong Kong, March 2004, pp. 80–91.
52. X. Wang, G. Xing, Y. Zang, C. Lu, R. Pless, and C. Gill, "Integrated coverage and connectivity configuration in wireless sensor networks," *ACM SenSys*, Los Angeles, November 2003.
53. G. Werner-Allen, K. Lorincz, M. Ruiz, O. Marcillo, J. Johnson, J. Lees, and M. Welsh, "Deploying a wireless sensor network on an active volcano," *IEEE Internet Computing*, vol. 10, no. 2, pp. 18–25, April 2006.
54. D.B. West, *Introduction to Graph Theory*, Prentice Hall, Upper Saddle River, New Jersey, 2001.
55. P. Yang, R.A. Freeman, G.J. Gordon, K.M. Lynch, S.S. Srinivasa, and R. Sukthankar, "Decentralized estimation and control of graph connectivity for mobile sensor networks," *Automatica*, vol. 46, pp. 390–396, 2010.
56. F. Yu, Y. Choi, S. Park, Y. Tian, and S. Kim, "A hole geometric modeling in wireless sensor networks," *International Conference on Wireless Communications, Networking and Mobile Computing*, pp. 2432–2435, September 2007.
57. H. Zhang and J.C. Hou, "Maintaining sensing coverage and connectivity in large sensor networks," *Wireless Ad Hoc Sensor Networks: An International Journal*, vol. 1 no. 1–2, pp. 89-124, 2005.

58. Y. Zou and K. Chakrabarty, "Sensor deployment and target localization based on virtual forces," *IEEE INFOCOM 2003*, March 2003.
59. Y. Zou and K. Chakrabarty, "Uncertainty-aware sensor deployment algorithms for surveillance applications," *Proc. IEEE Global Communications Conf*erence, December 2003.
60. Y. Zou and K. Chakrabarty, "Sensor deployment and target localization in distributed sensor networks," *IEEE Trans. Embedded Comput. Syst.*, vol. 3., no. 1, pp. 61–91, 2004.

Chapter 6
Localization and Tracking in WSNs

6.1 Introduction

Localization—the process by which the positions of the nodes of a Wireless Sensor Network (WSN) are found with respect to some absolute or relative frame of reference—is fundamental to how the WSN performs at executing its functions. Critical WSN operations such as routing (e.g., geographical routing), data aggregation and navigation (for mobile sensor networks) all heavily rely on the localization mechanism. For many of today's systems or applications requiring localization functionality, the NAVSTAR Global Positioning System (GPS) is typically sufficient [1]. In WSNs, however, GPS-based methods fall short for two main reasons. First, effective operation of a GPS-based localization mechanism demands that the WSN system has line-of-sight communication with multiple satellites. This requirement is typically not realizable because most WSN applications are meant for environments inherently having obstructions to electromagnetic signals, e.g., in urban areas (i.e., in the midst of tall buildings), indoors, under water, in forests or in mountainous areas to mention but a few. The second challenge posed by GPS-based techniques is their prohibitive price—the cost of the full network can increase over tenfold if a small subset of the nodes is equipped with GPS receivers [17]. With GPS being unsuitable for the vast majority of WSN applications, research on localization in WSNs is mostly focused on GPS-less techniques, which overcome the challenges seen with a GPS.

In this chapter, we examine some of these techniques. We first explore the localization scenario where all nodes are stationary (i.e., the layout problem [9]). We later extend our discussion to the case where node mobility is involved (i.e., where some WSN nodes are mobile) and present the algorithms and extra design considerations, which are prompted by node mobility. We wind up the chapter with a brief discussion on object tracking in WSNs, a very closely related problem to localization.

© Springer International Publishing AG 2016
R.R. Selmic et al., *Wireless Sensor Networks*,
DOI 10.1007/978-3-319-46769-6_6

6.2   Design and Evaluation of Localization Algorithms

To design or analyze the applicability of a localization algorithm to a given WSN application, one has to consider a range of factors, that include, the resource requirements of the algorithm, the topology of the network, the nature of the terrain in which the WSN will be deployed and the density of nodes in the network [2]. We discuss these factors in this sub-section.

Node Density The density of nodes in the WSN has a major bearing on what kinds of localization algorithms are suited for the network. For hop-count based algorithms for instance, a high density of nodes is required to ensure accuracy of the approximated distances [2]. Where beacon nodes are part of the localization process, their density must be high enough for the localization operations to be effective. In general, many localization algorithms demand a certain threshold node density below which the localization error may increase significantly but above which the error reduces only very slightly. As an example, in a beacon-driven localization simulation based on 500 WSN nodes deployed in a 100 m × 100 m × 100 m area [28], it was found, for an anchor (or beacon) percentage of 20 %, that the localization coverage increased by almost 50 % when the node density (which was represented as the expected number of nodes in a node's neighborhood) increased from 8 to 11, yet increased almost negligibly when the node density increased from 12 to 16.

Environmental Factors Obstacles such as buildings, rocks, and trees in the area where the WSN is deployed can impede signals used for the measurement of signal ranges (ranging methods discussed in Sect. 6.3.1.1) and result in an erratic localization process. For example, signals reflected by physical obstacles located within the WSN may interfere with each other, resulting into multipath effects and associated localization errors [1]. Besides physical obstacles, other environmental factors such as precipitation and the amount of moisture in the air are well known to affect radio wave propagation, potentially causing errors for localization techniques that rely on radio waves [1]. A good localization algorithm should have mechanisms to guard against or recover from the errors caused by environmental factors such as those listed here.

Network Topology Irregular WSN topologies typically result into higher localization errors [10]. Even where the topology may not be irregular, nodes at the edge of the WSN generally tend to be relatively difficult to localize since they: (1) have a small number of neighbors, and, (2) have all their neighbors on one side, which implies that range measurements for these nodes provide only a limited perspective of their location [2]. A good localization algorithm should be able to make the necessary compensations for errors resulting from topology artifacts.

Resource Constraints The design of localization algorithms for WSNs must be cognizant of the fact that WSN nodes have limited processing power and memory. For applications where low precision of localization measurements is adequate, approximate algorithms that can estimate position using low power and cheap hardware offer a good option around the resource limitations challenge [15]. When an application requires localization information at a high precision, the exact algorithms needed for this purpose generally consume more power. A good algorithm in such a case would have to distribute localization tasks between different nodes (e.g., tasks split between the base station, beacon nodes (which are typically more powerful than the typical node) and the rest of the nodes).

6.3   Categorization of Localization Approaches

In terms of the algorithmic methodology used to make localization computations, localization algorithms in WSNs are categorized as either range-based or range-free. Range-based methods use estimates of distance or angles to localize the WSN's nodes. From these measurements, simple geometric relationships are used to compute node locations without making assumptions about the underlying topology of the WSN. Range-free methods on the other hand rely on connectivity information in the WSN to estimate node locations. For effective localization, many range-free methods rely on the WSN's topology meeting certain requirements (e.g., that hop distances have low variance). The main advantage of range-free methods is that they do not require specialized hardware for distance and angle measurements, which makes them cost-effective in comparison to range-based methods. The key advantage of range-based methods on the other hand is the fact that they tend to be more accurate in comparison to the range-free methods [9]. Regarding the way in which localization computations are undertaken, certain (range-free or range-based) localization algorithms perform their localization computations in a distributed manner, while others operate in a centralized fashion. The latter approach has the core localization computations running on dedicated nodes (e.g., base station), while the former has the core localization operations running on the individual nodes.

In this section we explore some of the most popular localization schemes in the literature. We discuss several prominent range-based and range-free algorithms in details, and then finally give a general comparison between the centralized and distributed design approaches for WSN localization algorithms.

6.3.1   Range-Based Methods

These methods operate in two steps: a measurement step in which the distance/angle measurements are made, and a computation step in which the recorded measurements are combined to do the actual localization. The four main approaches used in the measurement step[1] include Received Signal Strength Indicator (RSSI), Time of Arrival (ToA), Time Difference of Arrival (TDoA) and Angle of Arrival (AoA) and are discussed next before we present the computation approaches used in the final localization step.

6.3.1.1 Approaches to Making Ranging Measurements

Received Signal Strength Indicator (RSSI) Theoretically, the power of a radio signal at a given point is known to be inversely proportional to the square of the distance of the point signal source. This relationship forms the backbone of the RSSI technique. If the power of the signal at the transmitting node is known, then, using this relationship, the receiving node can estimate its distance from the sending node. The main advantage of this approach is that it requires no dedicated hardware, i.e., it only requires the sensors to have a radio, which most WSN nodes are expected to have anyway. In practice, however, this approach is for several reasons susceptible to noise. Physical obstacles (e.g., walls, people, etc.) absorb and reflect the waves, while different environments impact the propagation of radio waves differently (e.g., radio waves propagate differently over asphalt than over grass [9]). With the many possible sources of error, radio wave measurements made in real settings rarely agree with the theoretical relationship between signal strength and distance traversed by the signal.

For example in [12], an experiment based on Intel's crossbow motes revealed that RSSI measurements taken from different directions of the sensors (e.g., north, east, west) did not depict a consistent relationship with the distance from the sensors. In this particular experiment, researchers minimized potential sources of noise, e.g., sensors had their batteries fully powered up throughout the experiment, the surface on which the experiments were done was level, no physical obstacles were present and electronic equipment that could potentially cause interference were not present in the vicinity of the experimental apparatus. With this closely controlled environment failing to demonstrate the power of the RSSI-based ranging method, the research raised doubts about how well RSSI would perform in a real deployment in the wild. Several other studies have made similar findings on the ineffectiveness of RSSI as a ranging method for WSNs. As of today, research on the development of reliable RSSI-based methods for WSN localization continues to be ongoing.

Time of Arrival (ToA) This method has two variants: One-Way Time-of-Arrival (OW-ToA) and Two-Way Time-of-Arrival (TW-ToA). In the OW-ToA approach, the sender and receiver of a signal have synchronized clocks. When the signal arrives at the receiver, it registers the time of arrival, and the time of transmission of the signal (this time is sent to the receiving node) and uses these two variables to

---

[1]We focus on these four main methods; however, there exist several derivatives of these methods in the literature.

compute the distance between the two nodes. The difference, t_{ij}, between the time of transmission of the signal at the sending node and the time of receipt at the receiving node can be obtained as (see [13]):

$$t_{ij} = \frac{\|x_i - x_j\|}{c} + e_{ij}, \tag{6.1}$$

where $\|x_i - x_j\|$ is the actual distance between the two nodes, c is the signal propagation speed (which should be known for the medium in question) and e_{ij} is the error term which follows the Gaussian distribution with zero mean and variance σ_{ij}. The distance estimate d_{ij} is obtained from,

$$d_{ij} = ct_{ij} = \|x_i - x_j\| + ce_{ij}. \tag{6.2}$$

In general, a high Signal-to-Noise Ratio helps minimize the estimation error. Three major challenges faced by this method are: (1) it suffers from unreliable measurements in the event that the clocks go out of sync, (2) extra communication overhead is incurred as each source has to send the time of signal transmission to the receiver, (3) since radio signals travel at the speed of light, recording their time of arrival precisely is a challenge [1].

The TW-ToA approach eliminates the need to synchronize clocks as the distance between the communicating nodes is estimated based on the round-trip delay of the signal. The sensor node i sends a signal to the receiver node j. After a turn-around time t_j^a the receiving node sends back a message to j to acknowledge receipt of the signal. Using a similar notation to that of the ToA method, the distance estimate d_{ij} can be expressed as:

$$d_{ij} = \|x_i - x_j\| + ct_i^a + c\left(\frac{e_{ij}}{2} + \frac{e_{ji}}{2}\right), \tag{6.3}$$

where e_{ij} and e_{ji} are the estimation errors at nodes j and nodes i for the signals being transmitted from them. This method eliminates the error due to imprecise synchronization of the clocks on nodes i and j; however, the method is also susceptible to errors if the clock in the reference node undergoes a drift.

Time Difference of Arrival (TDoA) This method computes the distance between two nodes based on the difference between the times of arrival of the radio signal and a second signal (typically an ultrasound or an audible frequency, which require each node to have a speaker and microphone [2]). The transmitter first sends a radio message, and then waits for an interval t before sending the sound wave. On receiving the radio signal, the receiving node switches on its microphone to detect the incoming audio signal. If the radio signal and audio signal are, respectively, received at times t_r and t_s, the distance d between the two nodes can be estimated from

$$d = (c - v)(t_s - t_r - t),$$ (6.4)

where c and v are, respectively, the propagation speed of sound and radio waves.

The TDoA concept can also be used with multiple sensors/receivers that are tasked with localizing a sensor node or a target in their sensing field. Consider a group of sensors shown in Fig. 6.1 and a hidden emitter that the WSN is trying to localize, i.e., consider a network of N sensors with coordinate vectors $\vec{p}_i(t)$, $i \in \{1, 2, \ldots, N\}$, and a hidden emitter located at $\vec{p}_e(t)$. Included are only sensors that are selected to localize the specific emitter source/sensor node. The distance between the emitter and the sensor i is $r_{ei}(t)$.

The TDoA sensing concept allows for measurement of difference in time of arrival, i.e.,

$$r_{ij}(t) = r_{ei}(t) - r_{ej}(t) = c(t_i - t_j),$$ (6.5)

where c is the speed of a signal propagation, $r_{ei}(t)$ is the distance between the emitter and the i-th sensor, and t_i is the signal propagation time between the emitter and the i-th sensor. Without a loss of generality, consider the coordinate system origin to be located at sensor 1. The closed form of TDoA-based localization is given in [18] using basic geometry of the sensor network

$$\left(P_e + r_{ij}\right)^2 = \|\vec{p}_i\|^2 - 2\vec{p}_i\vec{p}_e + P_e^2,$$ (6.6)

where P_e is the distance of the emitter from the coordinate system origin. This equation is equivalent to

$$0 = \|\vec{p}_i\|^2 - 2\vec{p}_i\vec{p}_e - r_{ij}^2 - 2P_e r_{ij}.$$ (6.7)

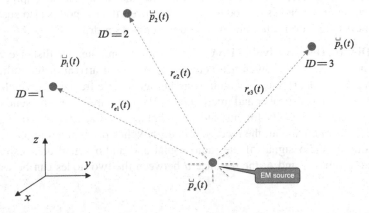

Fig. 6.1 Sensor network localize the hidden emitter using TDoA framework

However, the TDoA measurements will produce errors in the above equation, yielding

$$\varepsilon_i = \|\vec{p}_i\|^2 - 2\vec{p}_i\vec{p}_e - r_{ij}^2 - 2P_e r_{ij}, \qquad (6.8)$$

or in the matrix form [18]

$$\vec{\varepsilon} = \mathbf{P} - 2\mathbf{S}\vec{p}_e - 2P_e\mathbf{R}, \qquad (6.9)$$

where matrices are defined as:

$$\mathbf{P} = \begin{bmatrix} \|\vec{p}_2\|^2 - r_{21}^2 \\ \|\vec{p}_3\|^2 - r_{31}^2 \\ \vdots \\ \|\vec{p}_N\|^2 - r_{N1}^2 \end{bmatrix}, \quad \mathbf{S} = \begin{bmatrix} x_2\ y_2\ z_2 \\ x_3\ y_3\ z_3 \\ \vdots \\ x_N\ y_N\ z_N \end{bmatrix}, \quad \mathbf{R} = \begin{bmatrix} r_{21} \\ r_{31} \\ \vdots \\ r_{N1} \end{bmatrix} \qquad (6.10)$$

Note that the TDoA measurements will affect matrices $\mathbf{P}$ and $\mathbf{R}$, while matrix $\mathbf{S}$ contains locations of sensors. Then, the least-squares error solution for the optimal emitter location estimation is given by

$$\vec{p}_e = \frac{1}{2}(\mathbf{S}^T\mathbf{S})^{-1}\mathbf{S}^T(\mathbf{P} - 2P_e\mathbf{R}). \qquad (6.11)$$

This solution assumes that the distance of the emitter from the coordinate system origin P_e is known. To get an optimal solution in terms of $\vec{p}_e$ only, one needs to optimize the modified Eq. (6.9) in terms of P_e. The modified Eq. (6.9) in given by

$$\vec{\varepsilon} = \mathbf{P} - \mathbf{S}(\mathbf{S}^T\mathbf{S})^{-1}\mathbf{S}^T(\mathbf{P} - 2P_e\mathbf{R}) - 2P_e\mathbf{R}, \qquad (6.12)$$

$$\vec{\varepsilon} = (\mathbf{P} - 2P_e\mathbf{R})\left(\mathbf{I} - \mathbf{S}(\mathbf{S}^T\mathbf{S})^{-1}\mathbf{S}^T\right). \qquad (6.13)$$

The closed-form solution is obtained by minimizing $\vec{\varepsilon}^T\vec{\varepsilon}$ and is given by

$$\vec{p}_e = \frac{1}{2}(\mathbf{S}^T\mathbf{S})^{-1}\mathbf{S}^T\left(\mathbf{I} - \frac{\mathbf{R}\mathbf{R}^T\mathbf{S}_1\mathbf{S}_1}{\mathbf{R}^T\mathbf{S}_1\mathbf{S}_1\mathbf{R}}\right)\mathbf{P}, \qquad (6.14)$$

where $\mathbf{S}_1 = \mathbf{I} - \mathbf{S}(\mathbf{S}^T\mathbf{S})^{-1}\mathbf{S}^T$. It is recommended to first quickly calculate the emitter estimate and then to proceed with iterative methods for improved accuracy. Note that this method requires five or more sensors to accurately estimate the emitter location [18].

To minimize errors, TDoA requires that the media be free of echoes and the speakers be calibrated with the microphones since they tend to have different transmission and reception characteristics [2]. TDoA is considerably much more accurate than methods which entirely rely on radio waves. For the specific comparison between TDoA and RSSI, TDoA attains much better performance than RSSI because it only measures signal travel time yet RSSI measures signal magnitude. Signal magnitude measurements see noise from both occlusion and signal multipath effects, while signal time measurements only see noise from occlusion [2]. The major disadvantage of TDoA is the extra hardware it requires (e.g., microphones and speakers).

Angle of Arrival (AoA) This method uses several spatially separated radio or microphone arrays on the WSN node. When a WSN node receives a signal, differences between the phase of the signal at different microphones are used to determine the location of the transmitter. Increasing the number of array elements, the distance between them and the SNR helps improve the performance of the AoA method [13]. In a 2-dimensional setting without noise, a minimum of two receivers can be used to locate the transmitter. The presence of noise calls for the usage of more than two AoA measurements. The major challenge with the AoA technique is the expensive and bulky hardware (microphone and several speakers) it requires [2]. Moreover, the small form factor of the WSN nodes makes it difficult to accommodate multiple speakers that have enough separation as required for good performance.

6.3.1.2 Computing Locations from Ranging Measurements

From the angle and distance measurements, the most commonly used methods to find the locations of the WSN nodes are angulation, lateration, and statistical estimation. Angulation uses measured angles between nodes while lateration uses distance measurements between nodes to localize the nodes. For statistical estimation, the most commonly used techniques are Maximum Likelihood Estimation (MLE) and Bayesian inference. We next briefly discuss the mechanisms of these techniques.

Angulation This method is used when the angles or bearings of the nodes to be localized are known relative to the known locations of the anchor (or beacon) nodes (e.g., after application of the AoA technique). Triangulation is a specific form of angulation in which the angular separation between two anchors and the target node are used to localize the target node.

Figure 6.2 illustrates the triangulation mechanism. The two anchor nodes (Anchor #1 and Anchor #2) are at known positions and hence at a known distance, L, apart. Angles α and β represent the angular displacement of the target node from the two anchors. As illustrated in the Fig. 6.2, the meeting point of the two lines from the anchor determines the location of the target node. This location could for instance be expressed in terms of d, the perpendicular distance of the target node

Fig. 6.2 Illustration of triangulation

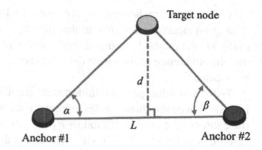

from the line joining the two anchors. Using simple trigonometry, d can be obtained using the following equation

$$d\left(\frac{1}{\tan \alpha} + \frac{1}{\tan \beta}\right) = L, \qquad (6.15)$$

which can be rewritten as

$$d = \frac{L \sin \alpha \sin \beta}{\sin(\alpha + \beta)}. \qquad (6.16)$$

In practice, the angular measurements α and β can be noisy, and the procedure can only define regions in which the target node is likely to be located. Angulation computations involving other nodes may then be used to fine-tune the position estimate.

Lateration This method is used when ranges between the target node and the anchor positions are known. Figure 6.3 illustrates trilateration, a form of lateration

Fig. 6.3 Illustration of trilateration

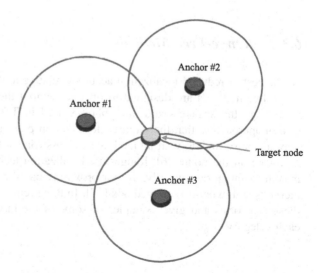

in which three anchor nodes are used to locate the target node. For instance, if ranging measurements reveal that the target node is a distance R_1 from the anchor node #1, the method stipulates that a circle of radius R_1 be drawn around #1, with the circumference of the circle defining the set of points where the target node could be located.

With a similar process undertaken for the other three anchors, the point of intersection of the three circles represents the location of the target node. Assuming the center of the circle with radius R_1 (i.e., the center of the circle around anchor #1) has the coordinates (x_1, y_1) with the centers and radii of the circles around anchors #2 and #3 defined similarly, the position (x, y) of the target node is found from the solution of the following three equations of the respective circles:

$$(x - x_i)^2 + (y - y_i)^2 = R_i^2, \quad i = 1, 2, 3. \tag{6.17}$$

Similarly as for angulation, measurement errors make it difficult to obtain the precise position of the target node. In such cases a region in which the target node is located is what is returned by the trilateration algorithm (as opposed to a precise point).

Estimation These methods use a measurement model expressing the relationship between the state of the system and measured data [1]. In Maximum Likelihood Estimation (MLE), the parameters capturing the system state are obtained by maximizing the likelihood of the measured data. The parameters are estimated using measured data with no prior information about state used. In Bayesian inference on the other hand, the system is estimated using both prior information and measured data. The estimation is based on recursive iteration, which use Bayes theorem [1].

6.3.2 Range-Free Methods

At the cost of reduced localization accuracy relative to the range-based techniques, range-free methods are designed to operate without the need for expensive hardware (e.g., the speakers and microphones used in TDoA). The idea behind this design approach is that the required localization precision for certain applications may not be so high to warrant the huge cost associated with the usage of expensive hardware on the nodes [6]. Range-free localization techniques can be generalized into three categories, namely, anchor proximity based methods, connectivity-based methods and event-driven methods [27]. In this section, we briefly discuss each of these categories and give examples of some of the most prominent algorithms in each category.

6.3.2.1 Anchor Proximity Based Methods

Localization under this approach is based on coarse-grained information of whether a given node is within the vicinity of another node. Based on a modality such as radio, infrared or sound, this localization approach uses binary information on whether a node A is within range of another node B, and then uses this information (in conjunction with similar information from other nodes) to carry out localization for the whole network.

The simplest example of an anchor proximity based localization method is the *Centroid* method [3, 27]. The method assumes a network in which a set of anchor nodes located at known positions (x_1, y_1) through (x_n, y_n) form a regular mesh and transmit signals containing their positions to the rest of the nodes. Each anchor node i is associated with a connectivity metric CM_i which is computed using

$$CM_i = \frac{N_{\text{rec}}(i, t)}{N_{\text{snt}}(i, t)} \times 100, \tag{6.18}$$

where $N_{\text{rec}}(i, t)$ is the number of beacons sent by i which have been received in time t, and $N_{\text{snt}}(i, t)$ is the number of beacons that have been sent by i in time t. Based on signals received from a subset of k anchors having CM_i exceeding a certain threshold CM_{th}, a node estimates its location, $(\hat{x}, \hat{y})$ as the centroid of the reference points, i.e.

$$(\hat{x}, \hat{y}) = \left(\frac{\sum_{\{j|CM_j \geq CM_{th}\}} x_j}{k}, \frac{\sum_{\{j|CM_j \geq CM_{th}\}} y_j}{k} \right). \tag{6.19}$$

To minimize the localization error, the method requires a dense network of anchors. Variants of this baseline Centroid localization algorithm incorporate additional heuristics, such as the use of weights to give more prominence to anchors closer to the node in question (see survey in [27]).

Another widely studied anchor-based algorithm is the Approximate Point in Triangle (APIT) algorithm [6]. This method segments the WSN into triangular regions whose vertices are the locations of anchor nodes. A node is localized based on the triangles to which it is found to belong. The method can be subdivided into three steps: (1) Beacon exchange—in this step nodes receive beacons from anchor nodes, (2) Point In Triangle (PIT) Testing—here a node chooses three anchors from all anchors from which it has received beacons and tests whether it is inside the triangle formed by connecting these anchors (this process repeats until all combinations are exhausted or the required accuracy is achieved), (3) APIT aggregation and centroid calculation—which involves the combination of results from different PIT tests to determine which triangle segments are more likely to contain a node, followed by a centroid computation which determines the location of the node.

Figure 6.4 illustrates how the results from multiple PIT tests are aggregated. A grid array is used to represent the area of the region that a node could occupy. The smaller the size of the grids, the better the accuracy. When a PIT test

Fig. 6.4 Scan algorithm for
PIT aggregation

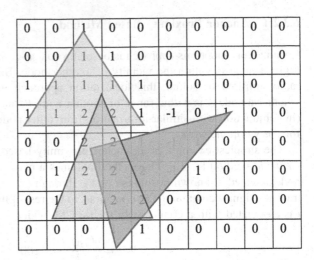

0	0	1	0	0	0	0	0	0	0
0	0	1	1	0	0	0	0	0	0
1	1	1	1	1	0	0	0	0	0
1	1	2	2	1	-1	0	1	0	0
0	0	2	2				0	0	0
0	1	2	2			1	0	0	0
0	1	1	2	2	0	0	0	0	0
0	0	0		1	0	0	0	0	0

determines that a node lies inside a given triangle, all cells in that triangle have their
score incremented. When the node is found to lie outside a given triangle, the scores
of the cells inside that triangle are decremented. At the end of the process, the
overlapping area with maximum score is used to calculate the centroid similar to
(6.19). This method also requires a high density of anchors for good performance.
Several variants of this method exist in the literature with a focus on attributes such
as anchor self-placement and optimization for WSNs with different properties [27].

6.3.2.2 Connectivity-Based Methods

Connectivity-based methods utilize connectivity information across the network to
make localization decisions. One of the most prominent amongst these methods is
the DV-hop method [11]. This method is centered on the distance vector routing
paradigm. Each anchor broadcasts a beacon that contains its location. The beacon
has its hop-count parameter initialized to one and incremented at each hop. As the
beacons from multiple anchors traverse the network, each node on their path registers the minimum hop-count value per anchor. Anchor nodes also keep track of
this information from beacons originating from their fellow anchors. If $dis(v_i, v_j)$
and $hop(v_i, v_j)$ denote the physical distance and a minimum number of hops
between anchors v_i and v_j respectively, the anchors estimate the average size of a
hop, D_{hop}, using

$$D_{hop} = \frac{\sum_{i \neq j} dis(v_i, v_j)}{\sum_{i \neq j} hop(v_i, v_j)}. \tag{6.20}$$

Using this information, an arbitrary node u_k can estimate its physical distance to the anchor v_i using

$$\text{dis}(u_k, v_i) = D_{\text{hop}}\text{hop}(u_k, v_i). \tag{6.21}$$

Based on information collected from multiple anchors, triangulation can be performed to localize a given node.

The challenge with this method is that the D_{hop} metric can only be representative of the actual per hop distance if the WSN topology is isotropic (i.e., if the physical distance of each hop is roughly constant in different directions). For networks having complex (anisotropic) shapes, the above formulations can produce very poor localization results. Several derivatives of the DV-hop algorithm exist in the literature, with some of them having mechanisms designed to tackle the irregular topology problem (see detailed survey in [27]).

The isometric feature mapping (isomap) algorithm also relies on sensor connectivity information for WSN localization [20, 27]. In this method, the number of hops, δ_{ij}, along the shortest path between two nodes in the WSN is used as an estimate of the actual distance d_{ij} between the two nodes. For a network containing n nodes, location estimation is done by minimizing the following cost function

$$C = \sum_{i=1}^{n} \sum_{j=1}^{n} \left(\delta_{ij}^2 - \|z_i - z_j\|^2 \right)^2, \tag{6.22}$$

where z_i is the estimated vector coordinates of node i and $\|z_i - z_j\|$ is the Euclidian distance between z_i and z_j. The optimal values of z_j are obtained using Multi-dimensional Scaling (MDS).

6.3.2.3 Event-Driven Methods

These methods use external localization events that are propagated through the WSN. The sensor nodes do not participate in the origination of the events. One of these techniques—the lighthouse method [16]—localizes a node based on the duration that the node dwells in a parallel rotating beam generated by the external localization device. The distance, d, between a target sensor node and the beam generator is estimated using

$$d = \frac{b}{2\sin(\omega \Delta t/2)}, \tag{6.23}$$

where ω is the angular velocity of rotation of the beam, b is the width of the beam and Δt is the interval at which the sensor node continuously senses the existence of illumination. The three-dimensional variant of this algorithm requires three mutually perpendicular beams of light.

Another localization algorithm in this family is Spotlight [19]. The algorithm largely follows the same mechanism as that of the lighthouse method, except for the fact that it moves all resource-intensive operations away from the WSN nodes and has them done on the external spotlight device. Several other methods using the same philosophy of localization have been proposed in the literature (see [27]).

6.4 Comparing Design Paradigms: Centralized vs. Distributed Techniques

A key question that has to be addressed before selecting a localization algorithm for a given application is whether the algorithm is centralized or distributed. Centralized algorithms have the distance/angle or connectivity information being sent from the nodes to a central processing center (e.g., the base station) where resource-intensive computations are carried out. Results from the computations are then sent back to the respective nodes [2]. Distributed algorithms have no dedicated computation unit and have all necessary computations done within the network (on both the anchor and regular nodes which engage in local information exchange). The main advantage of centralized algorithms is that they provide more accurate location information than their distributed counterparts. Their major disadvantages, however, are the lack of scalability (which makes them mostly suited for small scale WSNs) and the lower reliability arising from accumulated information losses seen with multi-hop transactions across a WSN [10].

In terms of communication energy efficiency, the difference between a centralized and distributed mechanism depends on the specific WSN setting. For a large network using a centralized scheme, the flow of localization traffic to and from the base station could cover a very large number of hops and hence results in significant energy usage. In a distributed setting, only local information exchange is carried out between neighboring nodes; however, many such exchanges may have to take place if a large number of iterations occur before a stable localization solution is obtained. The difference between the two varies depending on the specifics of the WSN application. For typical settings, past studies have found the distributed approach to be more energy efficient than the centralized approach when the number of iterations is less than the mean number of hops to the central processing unit [10, 14].

6.5 Localization in Mobile WSNs

6.5.1 Benefits of Node Mobility

When some of the nodes of a WSN are mobile, the WSN is said to be a Mobile WSN (MWSN). While mobility comes with increased energy consumption of the network, it has a number of advantages that include [1]:

(1) Network connectivity: In a static WSN, nodes in a certain part of the network can get completely disconnected due to battery drain. With the presence of mobile nodes, such connectivity issues are easily alleviated as the mobile nodes move to cover up for the connectivity gaps.

(2) Avoiding uneven node "death": Typically nodes at the edges of the WSN (towards the base station) die first because they handle most of the traffic that is being sent from the other WSN nodes to the base station. Through the use of mobile sinks, an energy consumption is more balanced across the network as all nodes take turns to forward data to the mobile sinks, which move towards these nodes at different points in time.

(3) Channel Capacity: The presence of mobile nodes enables multiple paths for data transport through the network. This increases the channel capacity and minimizes the likelihood that data integrity could be breached.

The simplest form a MWSN has, what is referred to as, the planar architecture. In this architecture, both the mobile and stationary nodes of the WSN communicate in an ad hoc manner over the same network [1]. In a 2-tier architecture the mobile nodes form an overly network or serve as "data mules" moving data through the network while in a 3-tier architecture the stationary sensor nodes pass data to the mobile nodes, which then pass the data over to the access points. Compared to static WSNs where localization is usually done only during the initialization stage, MWSNs require a continuous localization process as the nodes change positions in the network. This continuous localization presents new challenges, including localization latency and changes in the localization signal due to relative movement between the receiver and transmitter. We briefly describe these challenges next.

6.5.1.1   Algorithm Design Considerations Prompted by Node Mobility

Localization Latency Localization latency refers to that time interval between when measurements are made on a node and when the localization algorithms complete their computations to locate the position of the node. Given a mobile node in a WSN, the results of a localization computation are only meaningful if they are available soon after the measurements are done (i.e., when localization latency is kept to a bare minimum). If the localization algorithms take too long to render the localization decision, the node will likely have moved to a position far away from the previously computed position, resulting in erratic results for all other processes relying on localization information. Fast algorithms that overcome the localization latency problem tend to give less accurate localization results [5]. The design of a localization algorithm for a MWSN hence always has to make a trade-off between the localization latency and the accuracy of localization results. One common solution to the localization latency problem is the use of distributed algorithms that minimize the latency of data transmissions across the network [5].

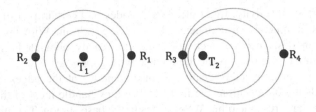

Fig. 6.5 Impact of the Doppler shift: no relative motion between transmitter and receivers (*left*); and transmitter and receivers moving relative to each other (*right*)

Doppler Effect Owing to the mobility of the transmitter, or the receiver, or both, the frequency of the signal as registered by the receiver may undergo a shift called the Doppler shift, which may in turn induce errors into the signal measurements fed into the localization algorithms. Figure 6.5 illustrates the Doppler effect where on the left there is no relative motion between the transmitter T_1 and the receivers, R_1 and R_2 (i.e., they could either all be stationary or moving at the same velocity). In this setting, the waves sent out by T_1 could be visualized as concentric rings which arrive at the receivers after fixed time intervals. To both R_1 and R_2, T_1 seems to be transmitting at a frequency determined by the rate at which the waves arrive at the receivers, which in turn is the actual frequency at which T_1 is indeed transmitting. The localization algorithms designed for the traditional static sensor networks are targeted towards this scenario and can reliably use frequency measurements made in this setting for their localization process.

Figure 6.5 on the right shows the situation in a MWSN where the transmitter T_2 moves relative to the receivers R_3 and R_4. With the transmitter moving towards R_3, each subsequent ring (assuming we visualize the signal as circular rings such as in the previous example) transmitted by T_2 arrives at R_3 faster than the previous one. Meanwhile at R_4, the reverse is true—as the signal takes longer and longer to arrive as T_2 moves away. For the receiver R_3, T_2 will appear to be transmitting at a certain frequency, while to R_4, it will appear to be transmitting at a different frequency. In truth, T_2 will not be transmitting at any of the two frequencies. This frequency shift caused by node mobility is what is referred to as the Doppler shift. For accurate localization in MWSNs, this shift has to be taken into consideration. The Doppler shift can be modeled using

$$\frac{\Delta f}{f} = -\frac{v}{c}, \tag{6.24}$$

where f is the frequency of the emitted signal, Δf is the frequency shift, c is the speed of signal propagation (speed of light for EM signals in air), and v is the speed of the source at which the source if moving away from the observer.

In practice a MWSN has a large number of nodes moving with varying velocities at different time instants. Compensating for the Doppler effect in a global localization framework hence requires the use of the above formulation while taking into consideration the movement properties of the network. An example of a localization model that compensates for the Doppler effect through elaborate modeling of the velocities and locations of the nodes can be found in [8].

Line of Sight Inconsistences In a MWSN, a node can have good line-of-sight communication with a mobile node at a given instant, and yet be in a position with a poor line-of-sight the next moment. This can negatively impact the localization process for mechanisms that rely on line-of-sight communication. This problem is generally addressed by having a high density of nodes around a given mobile node such that there are always a number of nodes in positions with good line-of-sight to the mobile node [1].

6.6 Tracking in WSNs

One of the application areas of WSNs is object tracking. Examples of such applications include, battle field surveillance (e.g., tracking of enemy tanks or soldiers in a battle field), tracking of animals in a forest and structural monitoring (i.e., monitoring structural response to forced excitation [22]) to mention but a few. In all these applications, the sensors have to initially detect the target, and then communicate amongst themselves to keep track of its position as it moves from one point to the next. A key aspect of this tracking process is how to efficiently detect the object and generate reliable reports in an energy efficient manner. There exists a wide range of tracking methods to address these issues in different ways. In this section, we briefly discuss the approaches to object tracking in WSNs. We make our presentation based on the three main families of tracking algorithms: namely, tree-based tracking, cluster-based tracking, and prediction-based tracking. The majority of all tracking algorithms borrow aspects from one or more of these algorithms, which implies that insights into their mechanisms should give a good picture of how tracking is done in WSNs in general.

6.6.1 Tree-Based Tracking

In this type of tracking, the network is modeled by a graph in which the vertices represent the WSN nodes, while the edges represent the connections between nodes that are able to communicate directly with each other. One of the most studied algorithms under this category is the Dynamic Convoy Tree Collaboration (DCTC) framework [26]. The method is centered on the idea of a *convoy tree*, which is a sub-tree of the full WSN tree which is comprised of the nodes around the moving target. When the target enters the WSN, the sensor nodes that first detect it select a root (which is usually a node which is closest to the target) amongst themselves and

Fig. 6.6 Mechanism of the
STUN algorithm

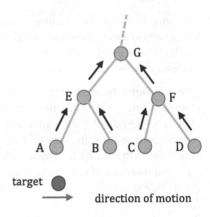

target ●

———▶   direction of motion

construct an initial convoy tree. The root collects more information from the nodes
so as to maintain a refined picture of the location of the target. As the target moves,
the *convoy tree* is modified with certain nodes far away from it being pruned, while
others are added to the tree. During this modification of the tree, the root node may
be replaced by another node that is located closer to the target. To minimize energy
usage during communication as the tree gets reconfigured with the movement of the
target, the DCTC is designed to always select a minimum cost convoy tree
sequence with high tree coverage. Selection of this tree is done through dynamic
programming performed on the optimization problem of finding the earlier mentioned minimum cost convoy tree.

In another method called Scalable Tracking Using Networked Sensors (STUN) [7],
a logical tree is built by successfully adding nodes to the tree based on the event rate
thresholds of the nodes. The tree is built using a bottom-up approach (from the leaves
to the root) with subsets of the sensors merged into balanced trees. Merging is done in
such a way that the high rate subsets are merged first. On this logical tree, the leaves act
as the sensors forwarding information up the tree. Figure 6.6 illustrates the operation
of this algorithm. As the target moves in the direction shown on the figure, the
closest leaf nodes A and B detect its presence. The two nodes will trigger their ancestor
E to register the target as a detected target, which will in turn alert its ancestor G about
the same information. As the target moves towards C and D, the two nodes will also
detect its presence and forward the message to their ancestor F, which in turn forward
the message to G. Because G will already be having this particular target among its
detected elements (after having been earlier notified by A and B), it will not forward
this message up the tree. Elimination of redundant message passing is central to
STUN's mechanisms for minimizing communication cost.

6.6.2 Cluster-Based Tracking

In cluster-based tracking, the WSN is segmented into clusters where each cluster
has a head node and member sensors. The distributed predictive tracking algorithm

in [25] is an example of a cluster-based tracking algorithm. The algorithm assumes a WSN that has already been segmented into static clusters. It distinguishes between sensors which are located at the border and those which are located deep in the WSN. Border sensors keep sensing at all times while the non-border sensors are in hibernation until notified by the cluster head to begin sensing. The idea behind this difference in operation of the border and non-border sensors is that the target of interest will originate from outside the WSN, and have to cross the border (and be sensed by the border sensors) before it can traverse the WSN. When a target is detected at the border, the Cluster Head (CH$_1$) for the group of sensors which first sense it formulates a unique descriptor for the target and sends it to the next downstream cluster head, (CH$_2$), and all the way to the sink.

The decision to send the message to CH$_2$ is based on a prediction step which determines that the most likely cluster head whose region is to be traversed next by the target is CH$_2$. This prediction is in turn based on the target's current speed and direction of motion at the time when it is detected by CH$_1$. Once CH$_2$ receives the message, it selects three sensors in its cluster that are closest to the predicted positions of the target and notifies them to "wake up" to sense the approaching target. This process continues through the network. In the event that the motion prediction step fails (e.g., if the target abruptly changes course), sensors within a *recapture* radius are all woken up to try to detect the target's new position. A key aspect of the algorithm's performance is its sensor hibernation mechanism which helps minimize its energy consumption. The main challenge with this method, however, is its static clustering approach (i.e., clusters are formed at time of net-work deployment and remain that way) which limits its tolerance to sensor faults.

Several dynamic clustering approaches have been proposed to address this drawback. In many of these methods, cluster formation is triggered by detection of the event of interest (see review in [4]) with no explicit CH selection needed (e.g., a sensor with sufficient battery power may volunteer to act as a CH). The algorithm presented for acoustic targets in [4] is an example of one such dynamic clustering approach.

6.6.3   Prediction-Based Tracking

Prediction-based tracking involves motion prediction steps that determine the likely destination of the target. This prediction helps with energy preservation as nodes which are far away from the region, that is predicted to be next visited by the target, can be put to sleep. Both cluster-based and tree-based algorithms can be designed to be prediction-based (e.g., see the Predictive Tracking algorithm discussed above). A key design attribute of prediction-based tracking is how the system recovers from prediction errors. Several papers in the literature propose different approaches to wake up the sensors once an error is detected (e.g., see [21, 23–25]) with one common criteria being minimizing recovery time and energy consumption.

Questions and Exercises

1. The Global Positioning System (GPS) is very widely used for the localization of objects in the earth's frame of reference. Why is GPS not suited for localization in WSN settings?
2. Time Difference of Arrival (TDoA) and Received Signal Strength Indicator (RSSI) are examples of ranging methods in range-based localization. Briefly describe the mechanisms behind the operation of these two techniques. Why does TDoA typically perform better that RSSI?
3. What are the advantages and disadvantages of range-free localization relative to range-based localization?
4. The DV-hop method is an example of a connectivity-based range-free localization method. Briefly describe how this method estimates the distance between an arbitrary node u_k and an anchor node v_i. Given distances of the arbitrary node from multiple anchors, describe how the node's location is determined through this method. Why does the DV-hop method fail in networks having complex shapes?
5. What benefits does the inclusion of mobile nodes bring to a WSN? Briefly describe possible different architectures of a WSN having some mobile nodes.
6. Briefly describe the meaning of the term "Doppler effect". How does this effect impact localization in mobile WSNs. How is the impact of this effect compensated for?
7. Briefly describe the mechanism of operation of tree-based, cluster-based and prediction-based tracking in WSNs.
8. During a WSN localization process, a target node is to be localized based on its angular displacement from two anchor nodes. Assuming the two anchors #1 and #2 are, respectively, located at the coordinates (2,3) and (10,0), and that the angular displacements α and β of the target node relative to the anchors #1 and #2 are, respectively, 45° and 60°, compute the location of the target node relative to the two anchors in Fig. 6.7.
9. In this problem you will use MATLAB to simulate the APIT localization algorithm. Assume that the WSN occupies a 20 × 20 region which is divided into 400 cells that are each 1 × 1 units in dimension. Let the bottom left corner

Fig. 6.7 Reference figure for
Question 8

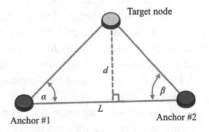

of the region have the coordinates (0,0) and the top right corner have the coordinates (20,20). Assume that the anchors at (0,0), (5,2) and (3,8) are connected to form a triangular region, just like the anchors at (10,10), (10,20), (15,10) and the anchors at (20,0), (15,5) and (15,0). Randomly generate 100 coordinates within this 20 × 20 region (assume the coordinates are integer numbers, i.e., each of the x and y coordinates are integer values between 0 and 20 inclusive). These coordinates represent the locations of the normal WSN nodes. For any five of these nodes that lie inside the triangles, use the APIT approach to find their locations. Compare the locations found by the algorithm to the actual locations of these nodes (compute the error as the Euclidian distance between the true locations and the computed locations and find the mean error over the five nodes). Rerun the APIT process when the WSN is segmented into cells that are 2 × 2 units in dimension and when they are cells that are 4 × 4 units in dimension. Comment on how the localization error varies in relation to the cell size (be sure to use the same 5 nodes in all three cases).

10. How many receiving localization nodes are enough for a successful implementation of a TDoA method?
11. What is the optimal configuration of receivers in TDoA? Please explain why.
12. Given four receivers in a plane that use RSSI method of localization, derive mathematically the optimal configuration of receivers. Simulate using MATLAB various scenarios and show that the solution found theoretically gives the best localization accuracy.
13. Describe how (6.11) can be used for localization by combining two different methods. Which method can be combined here? What is the trade-off in combining two methods versus using only one of the localization methods?

References

1. I. Amundson and X. Koustsoukos, "A Survey on Localization for Mobile Wireless Sensor Networks," in *MELT'09 Proceedings of the 2nd international conference on Mobile entity localization and tracking in GPS-less environments,* 2009.
2. J. Bachrach and C. Taylor, "Localization in Sensor Networks," in *Handbook of Sensor Networks: Algorithms and Architectures,* Hoboken, John Wiley & Sons, Inc., 2005.
3. N. Bulusu, J. Heidemann and D. Estrin, "GPS-less low-cost outdoor localization for very small devices," *IEEE Personal Communications,* vol. 7, no. 5, pp. 28–34, 2000.
4. W.P. Chen, J. Hou and L. Sha, "Dynamic clustering for acoustic target tracking in wireless sensor networks," *IEEE Transactions on Mobile Computing,* vol. 3, no. 3, pp. 258–271, 2004.
5. R. Fuller, "Tutorial on Location Determination by RF Means," *Proc. the 2nd International Workshop on Mobile Entity Localization and Tracking in GPS-less Environments, MELT,* Orlando, Florida, 2009.
6. T. He, C. Huang, B.M. Blum, J.A. Stankovic and T. Abdelzaher, "Range-free localization schemes for large scale sensor networks," *Proc. of the 9th International Conference on Mobile Computing and Networking,* New York, 2003.

7. H.T. Kung and D. Vlah, "Efficient Location Tracking Using Sensor Networks," *Proc. of 2003 IEEE Wireless Communications*, 2003.

8. B. Kusy, J. Sallai, G. Balogh, A. Ledeczi, V.-d. Protopopescu, J. Tolliver, F. DeNap and M. Parang, "Radio interferometric tracking of mobile wireless nodes," *Proc. of the 5th International Conference on Mobile Systems, Applications and Services*, 2007.

9. E.D. Manley, H.A. Nahas and J.S. Deogun, "Localization and Tracking in Sensor Systems," in *Proc. Sensor Networks, Ubiquitous, and Trustworthy Computing, International Conference*, 2006.

10. G. Mao, B. Fidan and B.D. Anderson, "Wireless Sensor Network Localization Techniques," *Computer Networks: The International Journal of Computer and Telecommunications Networking*, vol. 51, no. 10, pp. 2529–2553, 2007.

11. D. Niculescu and B. Nath, "Ad hoc positioning system (APS) using AOA," *Proc. 22nd Annual Joint Conference of the IEEE Computer and Communications*, 2003.

12. A.T. Parameswaran, M.I. Husain and S. Upadhyaya, "Is RSSI a reliable parameter in sensor localization algorithms: An experimental study," *Field Failure Data Analysis Workshop*, 2009.

13. M. R. Gholami, *Positioning Algorithms for Wireless Sensor Networks*, Gothenburg, Ph.D. Dissertation, Department of Signals and Systems, Chalmers University of Technology, 2011.

14. M. Rabbat and R. Nowak, "Distributed optimization in sensor networks," *Proc. the 3rd International Symposium on Information Processing in Sensor Networks*, 2004.

15. F. Reichenbach, J. Salzmann, D. Timmermann, A. Born and R. Bill, "DLS: A Resource-Aware Localization Algorithm with High Precision in Large Wireless Sensor Networks," *Proc. 4th Workshop on Positioning, Navigation and Communication*, Hannover, 2007.

16. K. Römer, "The Lighthouse Location System for Smart Dust," *Proc. the 1st International Conference on Mobile Systems, Applications and Services*, 2003.

17. M.L. Sichitiu and V. Ramadurai, *Localization of Wireless Sensor Networks with a Mobile Beacon*, 3 ed., Technical Report TR-03/06, 2003, pp. 174–183.

18. J.O. Smith and J.S. Abel, "Closed-form least-squares source location estimation from range-difference measurements," *IEEE Transactions on Acoustics, Speech, and Signal Processing*, vol. asp-35, no. 12, 1987.

19. R. Stoleru, T. He, J. A. Stankovic and D. Luebke, "A high-accuracy, low-cost localization system for wireless sensor networks," *Proc. the 3rd International Conference on Embedded Networked Sensor Systems*, 2005.

20. J.B. Tenenbaum, V.d. Silva and J.C. Langford, "A Global Geometric Framework for Nonlinear Dimensionality Reduction," *Science*, vol. 290, p. 2319, 2000.

21. Y. Xu and W.-C. Lee, "On localized prediction for power efficient object tracking in sensor networks," *Proc. 23rd International Conference on Distributed Computing Systems Workshops*, 2003.

22. N. Xu, S. Rangwala, K.K. Chintalapudi, D. Ganesan, A. Broad, R. Govindan and D. Estrin, "A wireless sensor network for structural monitoring," *Proc. of the 2nd International Conference on Embedded Networked Sensor Systems*, 2004.

23. Y. Xu, W. J. and W.-C. Lee, "Prediction-based strategies for energy saving in object tracking sensor networks," *Proc. International Conference in Mobile Data Management*, 2004.

24. Y. Xu, W. J. and W.-C. Lee, "Dual prediction-based reporting for object tracking sensor networks," *Proc. of the 1st International Conference on Mobile and Ubiquitous Systems: Networking and Services*, 2004.

25. H. Yang and B. Sikdar, "A protocol for tracking mobile targets using sensor networks," *Proc. of the 1st IEEE International Workshop in Sensor Network Protocols and Applications*, 2003.

26. W. Zhang and G. Cao, "DCTC: dynamic convoy tree-based collaboration for target tracking in sensor networks," *IEEE Transactions on Wireless Communications*, vol. 3, no. 5, pp. 1689–1701, 2004.
27. Z. Zhong, *Range-Free Localization and Tracking in Wireless Sensor Networks*, Ph.D. Dissertation, University of Minnesota, 2010.
28. Z. Zhou, J.-H. Cui and S. Zhou, "Localization for Large-Scale Underwater Sensor Networks," *UCONN CSE Technical Report: UbiNet-TR06-04*, 2006.

Chapter 7
Quality of Service

WSNs are used for a wide range of applications such as environmental sensing, clinical monitoring, and military surveillance. As in traditional data networks such as the Internet and Mobile Ad hoc Networks (MANETs), the different applications of WSNs demand that certain service requirements be met for optimal performance. The degree to which these requirements are met by a WSN is referred to as the Quality of Service (QoS) of WSN.

While Internet QoS is measured by the bandwidth, jitter, packet losses, and throughput, the notion of QoS in WSNs is dominated by a wider range of factors owing primarily to the resource-constrained nature of WSN nodes. In addition, WSNs have unique traits introduced by the factors such as unbalanced traffic (all flows leading from many nodes to a few nodes), data redundancy, and large numbers of nodes in a single network, all of which point to an additional set of QoS constraints relative to the Internet [4, 12, 21, 29, 30]. Resolving some of the constraints determining the QoS of WSNs often affects some other constraints, sometimes in a positive way and sometimes in an adverse way. A tradeoff decision is required based on the nature of the specific application when such effects are adverse.

In this chapter, we present details of QoS in WSNs, detailing the key QoS challenges in WSNs, and the different mechanisms that have been proposed to address these challenges.

7.1 QoS Building Blocks

Services rendered by WSNs include data and information gathering and assisting in timely and accurate decision-making with the gathered information. From the service standpoint, a QoS in WSNs is primarily determined by the quality of information and its timeliness. Since different resource constraints are major factors

© Springer International Publishing AG 2016
R.R. Selmic et al., *Wireless Sensor Networks*,
DOI 10.1007/978-3-319-46769-6_7

in WSNs, the quality of operation and quality of deployment also play significant roles in determining the overall QoS in WSNs [6, 10, 17].

The WSN QoS has two major perspectives: the network view and the applications view. Different applications have different demands regarding issues such as accuracy of sensor measurements, number of active sensors, total area of coverage, and more. This category of requirements represents the application-specific arm of WSN QoS. On the other hand, many WSN applications share a common range of network-specific demands, regarding issues such as timeliness of data delivery, data loss, and routing efficiency among others. These factors represent the second arm of WSN QoS. It is noteworthy though the boundary between network and application-specific QoS can sometimes be blurred. For example, the perceived accuracy of sensor measurements could be due to information loss during delivery by the network. Different QoS factors can also be tied to the different layers of OSI model, beginning with application to the Physical Layer. In this chapter, we present a unified view of WSN QoS without delving much into the sub-classifications (for detailed sub-classifications, see [4]). Below is a list of attributes that we refer to as QoS building blocks that determine WSN QoS.

Efficiency of Energy Utilization Given the resource-constrained nature of sensor nodes, energy efficiency is a very important aspect of QoS provisioning in WSNs. This requirement cuts across all WSN applications, and demands that all signaling protocols and the entire range of WSN algorithms must not be resource-intensive. One direct implication of this challenge is that WSN processes on the sensors must maintain as little network state as possible for energy to be used sparingly. If a single sensor node runs out of energy, the usefulness of the whole WSN may be affected since that node will not be able to detect and relay information from its location. Moreover, since WSNs are often deployed in hostile environments, it is usually infeasible to recharge or replace the batteries of a given set of nodes. In general, energy efficiency is the most important QoS determinant, since it directly controls the lifetime of the network, and the operation of the associated end-user applications.

Coverage WSN coverage is generally defined as the ratio of the space covered by the sensor nodes, to the total space of interest. The WSN applications require that every region of interest must be reachable by at least one sensor node. The existence of multiple coverage holes is a major cause of degraded quality of service to the applications that use the sensed data to make critical conclusions about the sensed environment (for more details see Chap. 5 on Coverage and Connectivity). Although a dense deployment of sensors will always result into a wide coverage area (with some areas even seeing redundant coverage), such deployment does not guarantee the complete coverage of the whole region of interest. As a matter of fact, the question of precisely determining the coverage area (or detecting coverage holes) of a WSN is still a major area of research, as it significantly impacts the QoS expected from a given WSN deployment.

Localization Accuracy In many WSN applications, a large number of sensors are deployed over a wide geographical area. For the sensed data to perform the intended task, it is important that these data be accurately associated with the precise location from which it originates. In addition, other processes such as routing and topology control demand that the locations of individual nodes be accurately known. The process by which node locations are determined is called localization. This process plays a major role in WSN QoS, since data that are associated with a wrong location can be misleading to the end-user of the particular WSN application. Manual configuration of each node with its location is infeasible in large WSNs having thousands of geographically distributed nodes. Another naïve approach to the localization problem is to equip each sensor with GPS, so that a centralized GPS satellite can accurately determine the position of each sensor. The challenge with this approach is that GPS demands line-of-sight communication, and is majorly hampered by obstacles like trees and buildings. In addition, GPS systems often impose a major hardware and resource burden on the small resource-constrained nodes, and can be a single point of failure of the entire WSN if attacked. These challenges have prompted research into alternative localization mechanisms that can guarantee WSN QoS under different adversarial conditions. An overview of the state-of-the-art approaches to localization is given in Sect. 7.2.2; a detailed treatment of the topic is done in Chap. 6.

Synchronization Most WSN applications require that all nodes have a common view of time. For instance, in a WSN deployed for vehicle tracking, different nodes may have to report the time at which the vehicle passes through a given set of coordinates, and then send this information to the sink node. Using these measurements, the sink node may then proceed to compute variables such as the speed of the vehicle. If the nodes do not have a common view of time, the computed speed will not be accurate, and the QoS perception of the end-user will be poor. Synchronization is the means by which all sensor nodes are set to use a common reference scale for their local clocks. In general WSN deployments, synchronization is also very useful in coordinating the sleep and wake-up events of the nodes so as to help save energy. Without well-coordinated scheduling of the above-mentioned events, a WSN may fail to detect important events. Time Division Multiple Access (TDMA) of the shared medium is another reason why WSNs must have a unified view of time. Unfortunately, synchronization is not a trivial problem for various reasons. First, it is a known fact that all hardware clocks are imperfect, since the oscillator frequency varies unpredictably due to various physical factors. As a direct result of this fact, local sensor clocks always drift away from each other with time. While protocols such as Network Time Protocol (NTP) suffice for Internet time synchronization, WSNs require simple schemes that do not impose significant resource burden on the sensors. A brief account of synchronization approaches in WSNs will be discussed in the next section.

Delays A number of WSN applications are event-driven delay-intolerant applications that demand a WSN to promptly report any detected events as soon as possible. On the other hand, query-driven applications demand that the required information is immediately available on request. In both types of applications, any delays in transmitting the observed signals may limit the usefulness of the relayed data. For such applications, the delay between detection of an event and propagation of the message to the end application is a very important determinant of WSN QoS. A simple example of a delay-sensitive, event-driven WSN application is a military target-tracking scheme that must immediately inform the sink about the temporal and spatial characteristics of the target so that it can be shot on the spot. Any data-transmission delays in such an application will render the delivered information useless.

Packet Loss In WSNs packet loss happens due to poor quality of wireless channels, failure of sensor nodes and congestion [7, 9]. Whereas some applications can tolerate some level of packet loss, other applications need reliable transmission of each packet. For either kind of application, unacceptable levels of packet loss may require re-transmissions of packets, draining the battery and reducing the WSNs' energy efficiency. Our focus in this chapter is on the congestion-related packet loss. Effective congestion control may require some level of (transport control) protocol management as well as load balancing. In the next section, we will briefly discuss how these can be performed.

7.2 QoS Provisioning in WSNs

In this section, we present a set of schemes explored in the literature for QoS provisioning in WSNs [2]. Each of these schemes addresses one or more the QoS blocks in WSNs described in the previous section, namely, energy efficiency, localization, synchronization, delay, and packet loss.

7.2.1 Topology Management

Techniques that seek to manage the sleep schedules of nodes in a WSN are generally referred to as topology management, since the active WSN topology changes every time a node sleeps or wakes up [11, 25]. Such techniques normally build on the fact that WSNs have high nodal densities, and that there is high spatial correlation of sensed data. To minimize the amount of energy consumed by a sensor node, it could be put to sleep (its radio turned off), and then woken up when there is data to be sent. Although this may result into increased latencies of data transfers (because a shortest-path node may be asleep), the approach can significantly increase the lifetime of the WSN if well implemented. Figure 7.1 shows a set of awake and asleep nodes (see left of the figure) with the resultant topology shown on the right. Note that the topology may change dynamically based on the number and

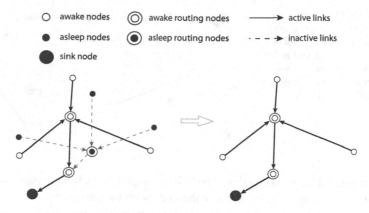

Fig. 7.1 Awake and asleep nodes with their possible links in a WSN (*left*) and the resultant topology for the awake nodes (*right*)

timing of the sleep and awake nodes. Only the awake nodes, including the awake routing nodes, are a part of the current topology.

A few of the proposed techniques for topology management are discussed below (note that some of the proposals overlap into generalized wireless ad hoc networks).

SPAN [5] According to SPAN, if a region has a high density of nodes, only a few of them need to be on at any time to forward data [5]. A small subset of nodes form a routing backbone, whose nodes see less sleep times than any other nodes. The backbone functionality is interchanged between different sets of nodes with time. The coordinators (backbones) stay awake continuously, while other nodes sleep and periodically check whether they should wake up and become coordinators. SPAN ensures that every node is within radio range of a coordinator. The algorithm also seeks to minimize the number of coordinators. The coordinator selection rule is such that if two nodes can only reach other via a certain non-coordinator node, then that (non-coordinator) node is eligible to be a coordinator.

Geographic Adaptive Fidelity (GAF) [31] Sensor nodes are divided into clusters such that every individual node within a cluster is seen as equivalent to all other nodes in the cluster (from the routing point of view) [31]. At no point in time is more than one single node active in a given cluster. GAF is primarily designed for mobile ad hoc networks; however, it is adaptable to WSNs as well. It identifies equivalent nodes between the source and the destination and use one node from each equivalent class at a time.

A diagrammatic representation of GAF is given in Fig. 7.2. GAF works in two stages. In the first stage the coverage area is divided into a set of square virtual grids and each node associates itself to the grid it is located in using GPS information. Two nodes are considered to be equivalent if they receive data from the sets of nodes and transmit data to the same sets of node. This ensures that if only one node from a set of equivalent nodes is awake and the rest are asleep, thus the routing is not interrupted.

Fig. 7.2 The GAF scheme: a
specific solution to the energy
efficient routing where a set of
nodes form a virtual grid and
in each such grid one sensor
node monitors and forwards
data while others are *left*
asleep

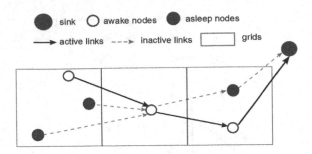

In the second stage, equivalent nodes in each grid, switch their status between
two states: asleep and awake. Usually the node with the maximum residual energy
becomes the master node and takes over the responsibility of forwarding data
toward the sink until it reaches a threshold energy level. Then it switches to sleep
mode and some other node with the maximum residual energy wakes up and act as
the master. In this way GAF can route data in an energy-efficient way [22] through
the management of redundant nodes.

Sparse Topology and Energy Management (STEM) [26] The motivation
behind STEM is that techniques such as SPAN and GAF can only produce significant benefits if the sensor nodes are very densely located [26]. STEM can work
for sparse networks, and even attains much more energy efficient if the network is
dense. The idea behind STEM is that a node should wake up only if it has to send
data. Since nodes spend most of their time waiting for events to happen, a lot of
energy can be saved if nodes can only awaken when there is indeed data to send.
Note that the sleeping of a node literally means that only the sensor's radio is off.
Such a node still has its processor running, and can still detect events. The processor generally consumes much less energy than the radio. The challenge under
STEM can be formulated as follows: if a node detects an event, how does it forward
data to its next hop neighbors whose radios are off? STEM approaches this challenge by having each node periodically turn on its radio for a short time to listen if
any other node needs to communicate with it. The node with data to be sent sends
out beacons with the ID of its next hop node to the sink. When the next hop node
receives this beacon, it turns on its radio to receive the data. The cycle then
continues until the data reach the sink. STEM uses a different frequency band for
the data from that of the wake-up packets to avoid interference effects.

7.2.2 Localization Techniques

In GPS-enabled sensor nodes the localization quality depends on the performance
of the localization accuracy of the GPS coordinate locator as well as the quality of
the link between a sensor node and the GPS coordinate locator. For sensor nodes
without GPS, a number of localization techniques have been proposed for WSNs.

These techniques generally operate in two phases: distance (angle) estimation and distance (angle) combination. The distance (angle) combination phase involves deterministic solution to some equations and hence is not a core QoS factor. However, the accuracy of distance (angle) estimation phase largely depends on the WSNs topology and is QoS factor of interest. Techniques used for distance estimation include Received Signal Strength Indicator (RSSI), Time-of-Arrival (ToA), and Angle-of-Arrival (AoA), and they are described in detail in Chap. 6 on *Localization and Tracking*.

7.2.3 Data Aggregation

Data aggregation techniques have been proposed to save WSN energy by filtering out redundant data from different sources so that it gets routed as fewer flows. This helps minimize the number of transmissions required to send the data to the sink, which in turn reduces the total energy expended by the network during the transfers. In addition, it reduces the amount of bandwidth required for the data transfers. A lot of research in the area of data aggregation in WSNs has focused on how to address the tradeoff between the improved energy efficiency and the increased delays sparked off by the data aggregation processes. A simple example of how data aggregation may cause increased delays is when data from nearer sources may have to be held back at an intermediate node to be aggregated with data coming from sources that are further away from the specific intermediate node. The latency due to data aggregation is generally proportional to the number of hops between the sink and the furthest source.

Data aggregation generally exploits the fact that data from a given group of neighboring nodes are closely correlated, and seeks to combine it in such a way to minimize redundancies [18, 19]. When sources send data to the sink, the intermediate routing nodes can look at the data and combine packets whose data are closely related. These intermediate nodes are known as data aggregators. There may exist a data aggregator for a cluster of nodes, or there may exist a tree level structure where a set of aggregated data is further aggregated in a hierarchical manner. Below we briefly describe them.

7.2.3.1 Cluster-Based Data Aggregation

Low-Energy Adaptive Clustering Hierarchy (LEACH) and Hybrid Energy Efficient Distributed (HEED) are two clustering approaches that can be used for data aggregation. LEACH collects data using the TDMA protocol to avoid collisions. The collected data are then transferred from the nodes to the cluster heads where aggregation is performed [14]. HEED improves the network lifetime over LEACH by optimizing the selection of cluster heads [34].

LEACH is a protocol that depends on cluster-based data aggregation for efficient routing. It works in two stages. At the first stage, a set of sensor nodes are randomly selected as clusterheads. Selected clusterheads then notify all other nodes about their status so that the non-clusterhead nodes can determine which clusterhead they should communicate to. The nodes in a cluster take their turn as the clusterhead after a certain interval so that a clusterhead is not totally drained of its energy. If all the nodes do not have uniform levels of energy, this can lead to selecting a clusterhead which is struggling to survive, i.e., has drained almost all its energy and is not qualified to serve as a clusterhead. Residual energy based selection of clusterhead can resolve this problem.

In the second stage of LEACH, the clusterheads receive data from the nodes belonging to their respective clusters, aggregate them and forward them toward the base station. During continuous monitoring, LEACH uses TDMA/CDMA MAC to avoid inter-cluster and intra-cluster collisions by creating and broadcasting a TDMA schedule to all the nodes allocating a time slot to each of the nodes in the associated cluster.

As mentioned above, HEED uses a similar approach to LEACH, however, it selects the clusterheads based on a pair of hybrid parameters one of which is the residual energy of a node and the other is either the node's proximity to its neighbors or the number of the node within that node's transmission range. This consideration in selecting clusterheads makes HEED more energy efficient than LEACH.

Figure 7.3 gives a general diagrammatic representation of data aggregation methods. There are two clusters, with each cluster having an aggregator node and the clusters are connected through routing node(s).

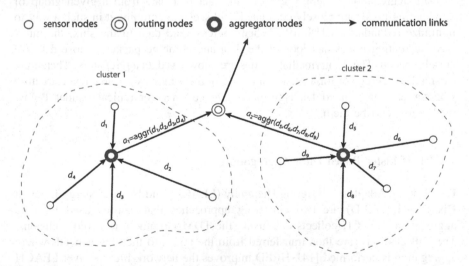

Fig. 7.3 Cluster-based data aggregation: an aggregator node aggregates data from a group of nodes (usually sensing similar information) and routes them through

7.2.3.2 Routing Tree-Based Data Aggregation

Tiny AGgregation service (TAG) and COUGAR are two of the initially routing tree-based approaches that support data aggregation [20, 32]. Using TAG and COUGAR, we can compute simple aggregated data, such as average temperature, from the sensors deployed to sense similar data from a common source using distributed SQL style query executed over a tree structure topology. Because of their intrinsic similarity, we briefly describe just one of them (i.e., COUGAR) below. We also discuss ACQUIRE, an improvement over COUGAR that can be used for efficient computation of complex queries [24]. Figure 7.4 shows a diagrammatic representation of routing tree-based data aggregation.

COUGAR introduces a query layer between the network layer and the application layer to initiate and propagate queries down through a tree structure and aggregate the data transmitted from the responding sensors. A predefined architecture of the sensor database system is used to generate query plans by the sink node that dictates each node determining the appropriate aggregator node for its data. This, however, requires the aggregator node to be more energy efficient than the others; otherwise the relevant nodes should take turn as the aggregator node. Also to avoid congestion at the aggregator node some sort of synchronization is also necessary.

In COUGAR, only the sensor nodes monitoring the events relevant to a query respond to a query. This on-demand data transmission combined with the data aggregation can significantly reduce the rate of energy consumption in a WSN.

ACQUIRE is a more efficient alternative to COUGAR. It divides a query into several subqueries and allows the intermediate nodes down to the source to respond to those subqueries, if they have updated data cached. If the cached data are not updated the query is propagated to the neighboring nodes until it is resolved. The query is replied with relevant data as soon as it is resolved. The sub-querying

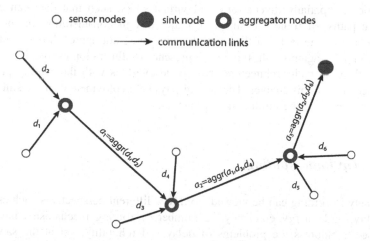

Fig. 7.4 Routing tree-based data aggregation: routing nodes usually serve as the data aggregators

allowing the intermediate nodes to respond make ACQUIRE much efficient than COUGAR, especially for the complex queries.

7.2.4 Load Balancing

In conventional WSN implementations, all nodes send data to a common sink. Therefore, sensor nodes closer to the sinks carry out a lot of packet forwarding and can have their energy drained much faster than the other nodes in the network. This in turn reduces the network lifetime. A second problem caused by a single centralized sink is the increased contention at the sink, and at the nodes closest to the sink (as all transfers head towards the same group of nodes), which can result into link overload and packet losses. This problem is further worsened by the fact that WSN traffic is normally carried in bursts, since nodes from a given location typically tend to respond to the same event, and as such try to send data at around the same time towards the sink. This increases the likelihood of overload whenever events occur. A few load-balancing architectures are briefly discussed below.

Load Balancing Through Clustering In this approach, independent clusters are defined in a WSN and in each cluster a less-energy-constrained gateway node is assigned to serve as a cluster head. The load balancing issue in a WSN is thus reduced to balancing loads among the gateway nodes only (see Fig. 7.5). The same concept can be extended to create clusters of gateways where a node in the gateway cluster serves as the cluster head. In this case it is important that each member gateway in a gateway cluster performs some sort of data aggregation/fusion operation so that the gateway cluster head is not overwhelmed with the amount of data to be forwarded.

Multi-sink Multi-path Architectures Multi-sink multi-path architectures provide a number of spatially diverse physical virtual sinks, such that data can traverse multiple paths en route the sinks. A major design consideration in multi-path architectures is the fact that the multiple paths must not approach one another, as contention may again result. Figure 7.6 presents an illustration of load balancing in multi-sink multi-path architecture. This issue overlaps with the routing problems described later in this chapter. The cost of physical deployment of more sinks is the major challenge of the multi-sink approaches.

7.2.5 Optimal Routing

Optimality in routing can be viewed from few different perspectives such as delay, reliability, and energy efficiency. A number of routing mechanisms have been proposed to address the problems of delay and reliability, yet at the same time

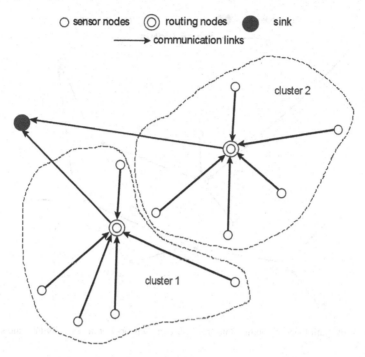

Fig. 7.5 Load balancing through clusters: a cluster head in each cluster routes the data for sensors residing in that cluster

utilizing resources efficiently [1]. Some major QoS-aware routing schemes are briefly touched on below.

Sequential Assignment Routing (SAR) [28] In SAR, reliability is ensured by having several multi-paths that have no common branches between a node and a sink [28]. Packet priorities are used, such that packets with higher priorities use paths with lower latencies. SAR creates multiple routing tree-based paths that lead to the sink. These paths avoid a node path suffering from critical energy level. A node may have access to several paths and selection of one or more paths by a node for routing is determined by the QoS requirement of the packet sent by that node and the cumulative QoS offered by a path. QoS requirement of a packet is usually determined by the priority level of that packet.

Reliability and fault tolerance, ensured by the access to multiple non-overlapping paths which are not prone to single route failure, is the prime contribution of SAR. It is important for SAR that route failure information is regularly updated so that an alternative route can be created. This is done periodically by the sink through verification of each individual path. The maintenance and monitoring of these paths add significant overhead to SAR.

Energy-Aware Data Centric Routing Algorithm (EAD) [3] The EAD seeks to eliminate redundant data by performing data pre-processing before forwarding

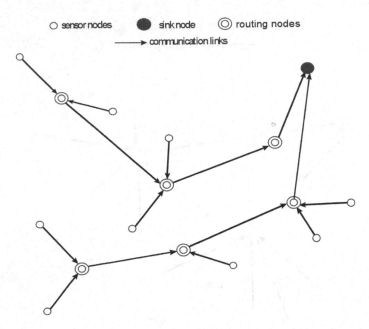

Fig. 7.6 Multi-path load balancing: data from different sensors not necessarily take the same path to the sink

data [3]. This class of routing algorithms falls into Directed Diffusion paradigm [16]. We have discussed some of these algorithms above while discussing data aggregation. Figure 7.7 shows an EAD with a query based routing method.

SPEED [13] SPEED is "*A Stateless Protocol for Real-Time Communication in Sensor Networks*", [13]. The SPEED protocol is designed to meet the strict minimal-delay requirements of real-time traffic. Under SPEED, a packet is forwarded through a neighboring node if the transmission rate via that node exceeds a set threshold. If no such neighbor can be found, SPEED drops the packet to help reduce the load in the network. The MMSPEED protocol seeks to improve SPEED's poor reliability, and uses priority queuing with the aid of multiple logical routing layers in the same physical network. Packets are routed through the different layers depending on the priority information contained in the respective packet headers. MMSPEED achieves reliability at the cost of higher resource consumption relative to the SPEED protocol.

Reliable Information Forwarding Using Multiple Paths (ReInForM) [8] ReInForM is another QoS-aware routing protocol that ensures reliable data transfer even in the presence of multiple errors [8]. ReInForM is built for large WSNs, since it requires a large topology with multiple routes. The algorithm makes multiple copies of each packet, (depending on the packet priority) and sends these copies along multiple links towards the sink. Depending on the importance of the data in packet, the channel error rate and the distance to the sink, the source computes the

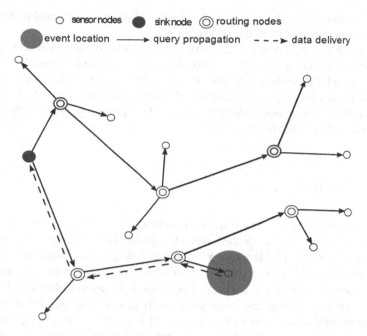

Fig. 7.7 Query-based routing supporting Energy-Aware Data Centric routing algorithm (EAD). A query is propagated from the sink down to the deployed nodes. Data is delivered only from the sensors monitoring the events involving the query

number of copies (or paths) required to deliver the packet with certain reliability. The source then sends the packet, and each node on the route uses the data in the packet header to make the next routing decision. Each intermediate node also updates the packet header fields. ReInForM does not require sensors to maintain network state as header information is used for routing decisions.

Minimum Cost Forwarding Algorithm (MCFA) [33] In the Minimum Cost Forwarding Algorithm (MCFA), sensor nodes do not maintain routing table, as they only aim to send data in the direction towards the sink. Each node maintains a record of its least cost distance to the sink, and every message sent out is broadcast to all neighbors [33]. If the node receiving a message checks and finds that it is on the least cost path to the sink, it re-broadcasts the message to its neighbors and the process continues recursively.

The least cost distance estimate in MCFA is done by broadcasting a message with so far incurred cost estimate (which is initially 0) from the sink to the neighboring nodes and rebroadcasting messages from the message receiving nodes. Each node upon receiving a message updates the cost estimate of delivering the message by adding the already incurred cost recorded in the message to the additional cost of receiving the message from the preceding node. To avoid conflicts due to multiple updates that may result in this process, each node is required to wait certain amount of time, known as back-off time, which is proportional to the cost of receiving the message from the preceding node.

Once all the nodes know the least cost estimate for delivering a packet to the sink from them, the routing decision is pretty simple. Each node broadcasts its data to its neighbors and the receiving nodes, if not a sink, make a decision whether to rebroadcast that message depending on whether it is located on the least cost path between the source sensor node and the sink. This process guarantees that MCFA delivers a packet with shortest delay.

7.2.6 Coverage

The coverage problem is addressed by positioning the sensors in such a way that the area of interest is comprehensively monitored [15, 23]. Several researchers have approached this problem by optimally positioning sensors such that a region is maximally covered. Other researchers have focused on finding a minimal number of sensors needed to monitor an area. There has been an established relationship between sensing coverage and node connectivity in sensor networks that states that if the communication range of the nodes is at least twice the sensing range then complete sensing coverage of a region implies that the network is connected. Below, we briefly describe few of the solutions to sensor coverage problem. Figure 7.8 shows an illustration of coverage. See Chap. 5 on Coverage and Connectivity for detailed analysis.

Coverage Via Subset Selection This approach assumes that the number of nodes deployed to monitor an area is higher than the required number. It then determines a

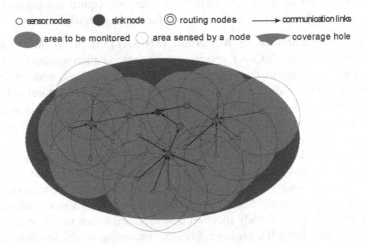

Fig. 7.8 Sensor coverage: the *gray shaded areya* shows how coverage may be missing (i.e., coverage holes) in a WSN network

minimal subset of the sensors that will maintain the desired level of coverage and powers off all other nodes to reduce power consumption.

Coverage Via Homology Using homology the absence of sensing coverage on a region known as coverage holes can be inferred using only local connectivity, without requiring positioning information. Sensors can be deployed to cover these regions known as coverage holes.

7.2.7 Synchronization

Synchronization helps coordinate sensor operations such as data aggregation and prolonging network lifetime through management of awake–sleep states. However, due to resource constraints, WSNs require synchronization schemes that do not require significant amount of memory or computing power. Few of the approaches of synchronization for WSN are briefly presented below.

Methods Based on Time Drifts These schemes determine the time drift at the nodes with respect to some reference nodes and then exchange those drifts and achieve synchronization by exchanging and comparing those drifts [27]. Solutions in this category include Reference Broadcast Synchronization (RBS), Timing-sync Protocol for Sensor Networks (TPSN), and Flooding Time Synchronization Protocol (FTSP).

RBS uses two-way message exchange to estimate drift and offset between clocks local to the sensor nodes. TPSN uses a tree to organize the network topology. The concept is broken up into two phases, the level discovery phase and the synchronization phase. The level discovery phase creates the hierarchical topology of the network in which each node is assigned a level. Only one node resides on level zero, the root node. In the synchronization phase all i level nodes will synchronize with $i - 1$ level nodes. This will synchronize all nodes with the root node. FTSP uses a structure similar to TPSN with a root node and that all nodes are synchronized to the root, but it improves on the disadvantages to TPSN.

Figure 7.9 shows an illustration of synchronization based on time drift. Synchronization nodes adjust for the drift.

Fig. 7.9 Synchronization based on time drifts: a reference node sends a time-stamped packet at time t and a receiving node computes its time drift by comparing the packet receive time t_i with t. Synchronization is achieved by exchanging the time drifts

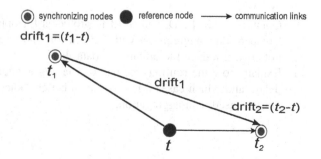

Coupling of OSI Layers Synchronization can be achieved by coupling the application layer with the MAC layer. This is done by time stamping a message with the time it is sent to the MAC layer. Thus, no central coordination for synchronization is required. This scheme has been shown to achieve synchronization up to within 2 μs.

Message Ordering This approach synchronizes clocks from the relative order of the messages received. Packets received from different sensors within a small time window satisfy this assumption if the broadcasting sensor nodes are distributed at about the same distance from the receiving node. Otherwise, there may exist significant timing drift between the sensors located at different distances and this simple scheme may not work.

Questions and Exercises

1. Some of the core QoS attributes of a WSN include: energy efficiency, coverage, localization, synchronization, delay and packet loss. Determine with justification which of these are attributes from the network view, application view, or both.
2. Explain how provisioning of each of the QoS blocks mentioned in question 1 may affect the other QoS blocks in adverse and helpful ways.
3. Name the factors that may introduce delay in WSNs. How can each of those issues addressed?
4. Why is redundancy almost inevitable in WSNs? What are the implications of such redundancies?
5. Compare multi-sink multi-path load balancing with cluster-based load balancing. Under what circumstances may the latter be considered as a variation of the former?
6. Is energy efficient routing always the best routing option? Please explain why or why not?
7. Some may argue that some QoS blocks, such as, localization accuracy largely depend on the algorithm deployed and should not be recognized as a QoS block. What are the strengths and weaknesses of this argument?
8. What are the positive and negative implications of having heterogeneous types of nodes, such as normal sensor nodes and energy enhanced aggregator nodes, in a WSN?
9. Compare SAR, EAD, and SPEED protocols for optimal routing in WSN.
10. Compare data aggregation with data fusion and explain why data fusion may not always reduce the amount of data flow.
11. Explain how the routing tree-based data aggregation approach may introduce delays and which cases it may yet be a better choice for data aggregation over cluster-based data aggregation.

12. Which synchronization issues may not be resolved if synchronization is achieved by coupling the application layer with MAC layer? Why?
13. What factors determine the optimality of routing in WSNs? Name a protocol to address each of those factors.

References

1. K. Akkaya and M. Younis, "An energy-aware QoS routing protocol for wireless sensor networks," *Proc. 23rd International Conference on Distributed Computing Systems Workshops*, pp. 710–715, May 2003.
2. V. Bahety and R. Pendse, "Scalable QoS provisioning for mobile networks using wireless sensors," *Proc. Wireless Communications and Networking Conference, 2004 WCNC*, vol. 3, pp. 1528–1533, March 2004.
3. A. Boukerche, X. Cheng, and J. Linus, "A performance evaluation of a novel energy-aware data-centric routing algorithm in wireless sensor networks," *Wireless Networks*, vol. 11, pp. 619–635, 2005.
4. D. Chen and P.K. Varshney, "QoS support in wireless sensor networks: A survey," *Proc. International Conference on Wireless Networks (ICWN 2004)*, pp. 227–233, Las Vegas, Nevada, USA, June 2004.
5. B. Chen, K. Jamieson, H. Balakrishnan, R. Morris, "Span: an energy-efficient coordination algorithm for topology maintenance in ad hoc wireless networks," *MobiCom 2001*, Rome, Italy, pp. 70–84, July 2001.
6. B. Deb, S. Bhatnagar, and B. Nath, "Information assurance in sensor networks," *Proc. the 2nd ACM International Conference on Wireless Sensor Networks and Applications*, pp. 160–168, New York, NY, USA, 2003.
7. B. Deb, S. Bhatnagar, and B. Nath, "Information assurance in sensor networks," Proc. the 2nd ACM International Conference on Wireless Sensor Networks and Applications, ACM Press, pp. 160–168, New York, NY, 2003.
8. B Deb, S. Bhatnagar, and B. Nath, "ReInForM: Reliable Information Forwarding Using Multiple Paths in Sensor Networks," *Proc. 28th IEEE International Conference on Local Computer Networks, LCN '03*. Bonn/Knigswinter, Germany, Oct. 2003.
9. R. DeRenesse, M. Ghassemian, V. Friderikos, and A.H. Aghvami, "Adaptive admission control for ad hoc and sensor networks providing quality of service," *Technical Report, Center for Telecommunications Research, Kings College London*, UK, 2005.
10. E. Felemban, C.-G. Lee, E. Ekici, R. Boder, and S. Vural, "Probabilistic QoS guarantee in reliability and timeliness domains in wireless sensor networks," *Proc. IEEE INFOCOM 2005*, Miami, FL, March 2005, vol. 4, pp. 2646–2657.
11. Y. Gao, K. Wu, and F. Li, "Analysis on the redundancy of wireless sensor networks," *Proc. the 2nd ACM International Conference on Wireless Sensor Networks and Applications, ACM Press*, pp. 108–114, New York, NY, USA, 2003.
12. A. Garcia-Macias, F. Rousseau, G. Berger-Sabbatel, L. Toumi, and A. Duda, "Quality of service and mobility for the wireless Internet," *Wireless Networks*, vol. 9, no. 4, 2003, pp. 341–352.
13. T. He, J.A. Stankovic, C. Lu, and T. Abdelzaher, "SPEED: A stateless protocol for real-time communication in sensor networks," *Proc. International Conference on Distributed Computing Systems*, Providence, RI, May 2003.
14. W. Heinzelman, A. Chandrakasan and H. Balakrishnan, "Energy-Efficient Communication Protocol for Wireless Microsensor Networks," *Proc. the 33rd Hawaii International Conference on System Sciences (HICSS '00)*, January 2000.

15. C.-F. Huang and Y.-C. Tseng, "The coverage problem in a wireless sensor network," *Proc. the 2nd ACM International Conference on Wireless Sensor Networks and Applications*, ACM Press, pp. 115–121, New York, NY, USA, 2003.

16. C. Intanagonwiwat, R. Govindan and D. Estrin, "Directed diffusion: a scalable and robust communication paradigm for sensor networks," *MobiCom 2000*, pp. 56–67, ACM New York, NY, 2000.

17. S. Kim, R. Fonseca, and D. Culler, "Reliable transfer on wireless sensor networks," *Proc. the 1st IEEE International Conference on Sensor and Ad hoc Communications and Networks (SECON04)*, Santa Clara, CA, October 2004, pp. 449–459.

18. B. Krishnamachari, D. Estrin, and S. Wicker, "Impact of data aggregation in wireless sensor networks," *International Workshop on Distributed Event-Based Systems*, Vienna, Austria, July 2002.

19. M. Kumar, L. Schwiebert, and M. Brockmeye, "Efficient data aggregation middleware for wireless sensor networks," *IEEE International Conference on Mobile Ad-hoc and Sensor Systems*, pp. 579–581, October 2004.

20. S.R. Madden, M.J. Franklin, J.M. Hellerstein, and W. Hong, "TAG: a tiny aggregation service for ad-hoc sensor networks," *Proc. 5th Annual Symposium on Operating Systems Design and Implementation*, Dec. 2002.

21. I. Mahadevan and K.M. Sivalingam, "Quality of service architectures for wireless networks: Intserv and diffserv models," *ISPAN*, pp. 420–425, 1999.

22. A. Mahapatra, "A QoS and energy aware routing for real-time traffic in wireless sensor networks," *Thesis/dissertation, University of Cincinnati*, Cincinnati, Ohio, November 2003.

23. S. Ramazani, J. Kanno, R. Selmic, and M. Brust, "Topological and Combinatorial Coverage Hole Detection in Coordinate-Free Wireless Sensor Networks," *International Journal of Sensor Networks*, vol. 21, no. 1, pp. 40–52, 2016.

24. N. Sadagopan, B. Krishnamachari, and A. Helmy, "The ACQUIRE mechanism for efficient querying in sensor networks," *Proc. the First International Workshop on Sensor Network Protocol and Applications*, Anchorage, Alaska, May 2003.

25. C. Schurgers, V. Tsiatsis, S. Ganeriwal, and M. Srivastava, "Topology management for sensor networks: Exploiting latency and density," *Proc. the 3rd ACM International Symposium on Mobile Ad Hoc Networking and Computing (MobiHoc)*, 2002.

26. C. Schurgers, V. Tsistsis, S. Ganeriwal, and M. Srivastava, "Optimizing sensor networks in the energy latency-density design space," *IEEE Transactions on Mobile Computing*, vol. 1, no. 1, pp. 70–80, 2002.

27. F. Sivrikaya and B. Yener, "Time synchronization in sensor networks: a survey," *IEEE Network*, vol. 18, pp. 45–50, July–Aug. 2004.

28. K. Sohrabi, J. Gao, V. Ailawadhi, and G.J. Pottie, "Protocols for self-organization of a wireless sensor network," *IEEE Personal Communications*, vol. 7, no. 5, pp. 16–27, Oct. 2000.

29. H.X. Tan, Quality of Service in Wireless Sensor Networks, Final Report in CS6204 - AY05/06 Semester II.

30. Y. Wang, X. Liu, and J. Yin, "Requirements of Quality of Service in Wireless Sensor Networks," *Proc. the International Conference on Networking, Systems, Mobile Communications and Learning technologies (ICNICONSMCL06)*, 2006.

31. Y. Xu, J. Heidemann, D. Estrin, "Geography-informed energy conservation for ad-hoc routing," *Proc. the Seventh Annual ACM/IEEE International Conference on Mobile Computing and Networking*, pp. 70–84, 2001.

32. Y. Yao and J. Gehrke, "The cougar approach to in-network query processing in sensor networks," *ACM SIGMOD Record*, September 2002.

33. F. Ye, A. Chen, S. Liu, and L. Zhang, "A scalable solution to minimum cost forwarding in large sensor networks," *Proc. the tenth International Conference on Computer Communications and Networks (ICCCN)*, pp. 304–309, 2001.

34. O. Younis and S. Fahmy, "HEED: A Hybrid, Energy-Efficient, Distributed Clustering Approach for Ad Hoc Sensor Networks," *IEEE Transactions on Mobile Computing*, pp. 366–379, October, 2004.

Chapter 8
WSN Platforms

8.1 Introduction

Previous chapters describe sensor network architecture, communication protocols and various characteristics related to security of WSNs. This chapter provides a short overview of hardware and simulation-based WSN platforms. A basic knowledge about the WSN platforms is necessary to understand physical and computational capabilities and limitations of the technology, to foresee future directions in technology development, and to further develop WSN algorithms and protocols.

While very few research and development results can directly be implemented on real wireless sensor nodes and networks, more often such implementations are first carried out on sensor network simulators. WSN simulators provide test platforms that can save time and cost, expand complexity of deployed networks, and thus verify results before any real hardware nodes are deployed. WSN simulators can verify results in simulation that can be costly, dangerous, or impossible to deploy in real-life, large-scale networks. Also, some applications in harsh environments would be extremely difficult for live testing, and effect on possible node failures is easier to simulate on WSN simulators. Therefore, WSN simulators are an integral part of every research and development effort in this area.

8.2 WSN Hardware Platforms

WSN platforms implement the physical layer of the protocol stack and have the primary goal of gathering multi-modal information about the physical phenomena. The three main components of a WSN hardware platform—namely, sensors which perform the sensing functions, the radio that provides communication and

© Springer International Publishing AG 2016
R.R. Selmic et al., *Wireless Sensor Networks*,
DOI 10.1007/978-3-319-46769-6_8

networking, and the microcontroller which performs the processing—are designed with an aim of minimizing the amount of consumed power. This design objective emanates from the fact that WSNs are usually deployed in environments where human intervention after deployment is not expected, a scenario that rules out the possibility of human-supported battery recharging.

One of the first wireless, low-power, real-time monitoring applications was developed by UC Berkeley and was called Mica motes [12]. Those first sensor nodes were based on low-power, 8-bit or 16-bit microcontrollers with few sensors either attached directly to the networking platform or with modular hardware approach where various sensing modules can be easily attached. Included are Mica and Mica2 wireless sensor networks, followed by IRIS platform, and Cricket system discussed later in the chapter. Since then, researchers are actively exploring WSNs applications in numerous domains and engineers have created more advanced platforms with improved computational capabilities [3] such as WSN nodes running on 32-bit ARM Cortex-M3 microcontrollers [3, 9] that reduces latency and energy consumption for computationally intensive tasks, but have increased energy consumption for traditional sensing applications. Further hardware platform development includes FPGA-based nodes [4, 10, 15, 16], that are modular and can be reconfigured. In particular, [15] presents a WSN node that combines FPGA and System-on-Chip technology and is a low-power, a high-performance, adaptable sensor node. Only recent FPGA could be used in such applications as their power consumption was reduced (initial FPGA were extremely power intensive devices).

In this chapter, we describe the most commonly used WSN hardware platforms that are important from historic/technology development perspective or have a high-market penetration. Table 8.1 shows the described WSN hardware nodes.

8.2.1   IRIS

The IRIS sensor nodes platform, from MEMSIC, Fig. 8.1, is an evolution of the first Mica and Mica2 sensor nodes platform [6, 26]. The nodes, also called Motes, run on a 2.4 GHz low-power Atmel's radio (AT86RF230) that is based on IEEE 802.15.4 standard [26] with transmission data rate of 250 kbps.

IRIS motes have an outdoor line-of-sight range of about 500 m and indoor range of about 50 m. They run on 2 AA batteries with current consumption ranging from 10 to 17 mA depending of transmission power of the radio. Motes are based on Atmel ATmega1281 8-bit microcontroller with 128 kB of Flash memory, 4 kB of EEPROM, and 8 kB is an internal SRAM. This is a low-power, static-operation, advanced RISC architecture microcontroller with selectable various frequency of operation from below 4 MHz and up to 16 MHz.

Table 8.1 Common WSN platforms with their microcontroller units, radio, and sensor components

WSN Node	MCU	Radio	Sensors
IRIS	Atmel ATmega1281, RISC architecture (Tiny OS)	2.4 GHz Atmel AT86RF230	Various sensor boards, communication supported using analog inputs, digital I/O, I2C, SPI, UART.
Digi XBee® ZigBee	No MCU, can communicate with variety of microcontrollers through serial interface	865 MHz to 2.4 GHz	No sensors on the module.
WiSense	TI MSP430 (Linux)	CC1000	Temperature, light, and other custom modules
Intel® Mote 2	Intel PXA271 (Tiny OS or Linux)	CC2420 at 2.4 GHz (IEEE 802.15.4)	Stackable modules, can support image and video processing
Mulle	ARM Cortex-M4 (Contiki OS)	IEEE 802.15.4 at 868 MHz, 900 MHz	Different I/O with 60-pin connector, various sensors can be added
CM30x	Jennic JN5148 RISC wireless microcontroller	IEEE 802.15.4 integrated with MCU	Multiple I/O with 34-pin connector for sensor modules
Fleck3	Atmel ATmega128 (Tiny OS)	Nordic nRF905, ISM band 433/868/915 MHz	Built in temperature and light sensors, can add an expansion board with extra sensors
Cricket	Atmel ATmega128L (Tiny OS)	CC1000	51-pin connector for sensor module
Shimmer wireless node	TI MSP430	TI CC2420 at 2.4 GHz (IEEE 802.15.4)	Sensor modules: 9DoF kinematic, GPS, ECG, etc.
ADVANTICSYS XM1000	TI MSP430	TI CC2420 at 2.4 GHz (IEEE 802.15.4)	Integrated temperature, humidity, and two light sensors

Fig. 8.1 IRIS sensor node from MEMSIC (reproduced by permission of MEMSIC, www.memsic.com)

Motes have 51-pin expansion connector that enables stacking of various sensor boards for different applications such as security and building monitoring, acoustic and vibration monitoring, chemical sensing, and more. Communications with multiple sensors is possible through analog inputs, digital I/O, I2C, SPI, and UART interfaces.

IRIS runs on TinyOS [28], an event-based operating system written in nesC programming language, which is a variation of a C language for embedded systems with small memory storage devices. An event handler processes events as they occur. The operating system consists of components connected with interfaces for various wireless sensor network abstractions such as communication, routing, sensing, and more. Components include the specification with names of their interfaces and the interfaces' implementation. Interfaces have a bidirectional feature that has been specified into a set of commands and a set of events. Components have two forms: modules and configurations. Modules can provide the application code that contains different interfaces. The modules specify what interfaces have been used or provided, and then implement them with the corresponding code. Configurations are used for assembling or linking other components together.

A typical application with IRIS WSN nodes is given in [24] where a passive radar is proposed that monitors for presence of people in indoor environments. The WSN nodes measure the Received Signal Strength Indicator (RSSI) that determines how people's motion affects RSSI at the pre-deployed WSN. The method is based on IRIS node capability to directly measure the received signal strength through its Atmel radio chip. In this case the radio serves dual purpose— provides wireless communication platform and measures received signal strength. In [19], IRIS nodes were used for ambient monitoring application as well as to verify energy modeling in WSNs.

8.2.2 WiSense

WiSense is a low-power, modular WSN platform that is based on the TI MSP430 microprocessor and CC1000 radio technologies operating in sub-GHz frequency range (see Fig. 8.2). The microcontroller, radio, and sensor components are all separated as different hardware modules (boards), with the microcontroller connecting to the radio through the SPI interface and to the sensors through I2C, SPI, and UART interfaces. The microprocessor board is equipped with temperature and light sensors as well as EEPROM. The platform is suitable for fast prototyping in home automation, smart building, and industrial automation applications.

The sensor node runs open-source Linux that implements IEEE 802.15.4 standard. The software and sensor nodes support both Full Function Device (can operate as a router node and can also transmit data from other nodes, thus supporting various network topologies) and Reduced Function Device (only sends information to network, cannot relay data from other nodes) modes.

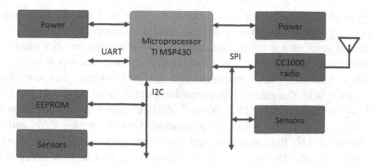

Fig. 8.2 WiSense wireless sensor network node block diagram with main design components

8.2.3 Digi XBee® ZigBee

Digi XBee® ZigBee 802.15.4 RF communication modules are a common platform for wireless connections between various electronic devices, not just sensor nodes, (see Fig. 8.3, [27]). They are a popular platform for wireless sensor networks where sensors can be added as separate components. They operate on frequencies from 865 MHz to 2.4 GHz and support various wireless interfaces including mesh, point-to-point, star, ZigBee, WiFi and others.

Digi XBee® ZigBee operates of 250 kbps data rate and has a range of up to 100 m for 1 mW transmit power devices or 1.6 km for 60 mW transmit power devices. The system supports direct sequence spread spectrum modulation and the 128-bit Advanced Encryption Standard for data encryption. However, MAC-layer addresses are non-encrypted and can be visible to all. The wireless module can operate in a command mode (node firmware can be modified through a set of commands, characters), an idle mode (listens for valid data or RF and serial ports), a receive mode (when a destination node receives a dedicated packet), a transmit mode (prepares to send packets as serial data), and a sleep mode (low-power state when node is not in use, cannot send or receive packets until awaken). Basic commands and changes in the Digi XBee® ZigBee node configuration and firmware can be entered remotely.

Fig. 8.3 Digi XBee® ZigBee wireless communication platform (reproduced by permission of Digi XBee®, http://www.digi.com)

In a typical application for condition monitoring and energy management in homes [7], Digi XBee® ZigBee nodes are used as Internet of Things (IoT) where three different sensors are connected with the coordinator/controller module using ZigBee communication standard in a mesh topology or a star topology (depending of the range between the coordinator and the end nodes). The coordinator is communicating with the gateway that translates the ZigBee protocol to the Internet protocol (IPv6) format. The Digi XBee® ZigBee sensor nodes produce ZigBee packets (64 bits address for a node on a Personal Area Network—PAN and a 16 bit address for the PAN) that are converted into IPv6 packets (128 bits to address a node on the network: 48 bits to address the network, 16 bits to address the PAN, and 64 bits to address the sensor node) and then sent to a central server. In the opposite direction, command packets towards the Digi XBee® ZigBee modules are encapsulated in a User Datagram Protocol (UDP) and then converted back to ZigBee packets by the IoT application gateway. This translation allowed for IoT IPv6 implementation of sensor nodes in 802.15.4 data format and ZigBee network where each sensor node is addressable with its specific IP address [7].

8.2.4 Intel® Mote 2

Intel® Mote 2 or iMote2 [29], from Intel®, is a high-performance WSN node that uses Intel Xscale® PXA271 CPU. It is able to operate at low power since the PXA271 CPU is capable to run at low voltage (0.85 V) and low frequency (13 MHz). The processor has several low-power modes including the sleep and deep sleep modes. In the deep sleep mode it draws a current of about 390 μA (compared to up to 66 mA in active mode). The processor integrates many I/O options that include I2C, 3 Synchronous Serial Ports, 3 high-speed UARTs, fast infrared, camera interface USB client and host and I2S codec audio interfaces among several others. These many I/O options make it very flexible in supporting different sensors. The processor also adds several timers and a real-time clock.

iMote2 integrates the IEEE 802.15.4 radio transceiver (CC2420). This transceiver supports a data rate of 250 kb/s with 16 channels in the 2.4 GHz band. Also, this mote platform integrates a 2.4 GHz surface mount antenna, which has a nominal range of up to 30 m. For applications requiring a longer range, an SMA connector can be soldered directly to the board for the connection of an external antenna. The module includes a power management IC that supplies the processor with the various voltage domains.

For interfacing the sensor board, the iMote2 has two sets of connectors, the basic set and the advanced set. The basic set supports the most common sensor interfaces and can be supported in future mote designs. The advanced set is platform-specific and exposes advanced features such as the camera interface, high-speed bus and audio interfaces. The mote can be powered by primary battery, rechargeable battery, or using mini-B USB connector.

Dialog DA9030 Power
Management IC

2.4 GHz 802.15.4
antenna

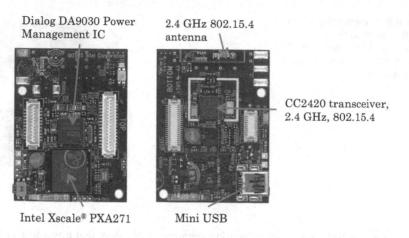

CC2420 transceiver,
2.4 GHz, 802.15.4

Intel Xscale® PXA271   Mini USB

Fig. 8.4 Intel® Mote 2 sensor node, *top view* (*left*) and *bottom view* (*right*) (reproduced by permission from [29])

The iMote2 supports a range of operating systems that include TinyOS for extremely low-power sensor network applications and Linux for more advanced applications. Detailed specifications of this node can be found in [29]. The following figures show the top and bottom of the Intel Mote 2 sensor node (Fig. 8.4).

8.2.5 Mulle

Mulle [30] (from Eistec AB, Fig. 8.5) is a wireless Embedded Internet System (EIS) for wireless sensors connected to the Internet of Things (IoT). Mulle platform comprises the Mulle sensor nodes, the Mulle Internet gateway device and the cloud services and Mulle development tools. The platform uses an ARM Cortex-M4 microcontroller and an IEEE 802.15.4 radio operating at 868 MHz. A frequency of 900 MHz is supported upon request. The Mulle is able to store large amounts of sensor and configuration data on its on-board 2 MB flash memory. It has a high-density expansion port with a 60-pin connector that supports a large number of I/O options (both digital and analog), enabling connectivity to other types of sensors and various debugging and programming tools. A selection of supported expansion boards includes [30]:

- Programming board, including a JTAG programmer and pin headers for all expansion port pins.
- IMU board equipped with gyro and magnetometer.
- Weather station board including sensors for barometric ressure, humidity, temperature, ambient light.
- Gateway board for using a Mulle board as a 6LoWPAN/RPL border gateway.

Fig. 8.5 Current version of a
Mulle sensor node
(reproduced by permission of
Eistec AB, http://www.eistec.
se/mulle/wsn)

Mulle runs the open-source Contiki operating system that features a full TCP/IP stack (with support for IPv6) and runs protocols such as UDP, TCP, HTTP and 6LoWPAN among others. The use of TCP/IP over 6LoWPAN enables the Mulle to transmit sensor data directly to the Internet. For Internet operations involving complex negotiations that cause high battery consumption, the gateway acts as a mediator for Internet services, with the aim to reduce the power consumption of the sensors. In its lowest sleep mode, the Mulle's power consumption is 20 μW. Mulle software can be built using the standard GNU GCC tool chain for embedded ARM platforms.

The legacy Mulle has a Renesas M16C/62P microcontroller and either a Bluetooth 2.0 module (v3.1, v3.2, v4.1) or IEEE 802.15.4 transceiver. Similarly as the new version, it also has an on-board 2 MB flash memory.

8.2.6   iSense Core Module 3 (CM30x)

The CM30x [31] uses the 32-bit Jennic JN5148 RISC wireless microcontroller that integrates the radio module in the single chip together with the microcontroller. It has 128 kB of memory that may be shared between program code and data. This flexibility of memory allocation enables a more robust operation than in designs in which the memory allocation is fixed. Its IEEE 802.15.4-compliant radio achieves a data rate of 250 kBit/s and supports two additional modes of operation, offering increased data rates of 500 and 667 kBit/s. The CM30x supports AES encryption, is ZigBee-ready and supports time of flight-based ranging.

There are three variants of CM30x, which differ in the antenna used: the CM30I uses an integrated PCB antenna, the CM30U uses a U.FL connector for external antenna, and the CM30HP uses a power amplifier and U.FL connector for external antenna. The module with the integrated PCB antenna (CM30I) is particularly well suited for compact systems. For the first two antenna options, a receive sensitivity of −95 dBm (at 250 kBit/s) and a transmit power tunable between −60 and +2.5 dBm is supported while for the third option, a receive sensitivity of −98 dBm (at 250 kBit/s) and a transmit power of up to 10 dBm is supported. The module has a LED that aids in debugging operations and a high-precision clock that with infrequent resynchronization enables precisely timed sleep and wake-up periods. A 34-pin connector that can supply up to 500 mA allows for connection to other modules (such as sensor modules, a gateway module, or an I/O module) to the core module. The core module may be powered by a wall-mount adapter, a standard battery holder, one of the power modules, or via an USB interface. The system has a voltage regulator that is software controlled.

8.2.7 Fleck3

The Fleck3 is the second generation of the Fleck series [2, 18], Fig. 8.6. Developed at the CSIRO ICT center in Australia, the Fleck series was designed to overcome some of the limitations of the Mica mote. Areas in which this series overcome the challenges of the Mica2 motes include: the provision of a one-board solution (screw

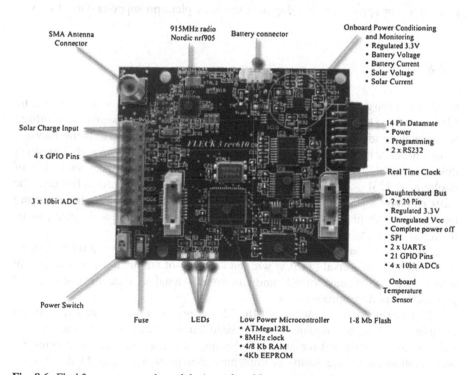

Fig. 8.6 Fleck3 sensor network module (reproduced by permission from [2])

terminals give access to digital and analog channels) and a robust expansion board interface, support for solar charging, and the use of a long-range radio having a range of over 1000 m.

The Fleck3 uses an Atmel ATmega128 microcontroller and a Nordic nRF905 radio. Communication between the radio and the microcontroller is via the SPI bus, for which the ATmega128 acts as the SPI master and the radio as an SPI slave device. The bus is central to the communication mechanism as units such as the flash memory, the real-time clock and the temperature sensor all communicate with the ATmega128 over the SPI bus. The radio works in all the different ISM bands (433, 868, 915 MHz) which is more suited for its main application area (e.g., outdoor environmental monitoring and agriculture applications) than the 2.4 GHz band used by wide-band radios following the IEEE 802.15.4 standard. The radio further has all its components integrated within the chip, which reduces the number of parts and the variability in radio characteristics.

The Fleck3 uses the DS1306 real-time clock (RTC) chip, which is interfaced to the Atmega128 as a SPI slave device. The clock uses <1 μA current for time keeping and <1 mA when active. The RTC frees up the sensor node from the need to keep time, which helps reserve resources for event-based programs. For security, an optional board that uses an off-the-shelf Trusted Platform Computing (TPM) chip from Atmel is supported. This chip implements the full TPM specification that includes a crypto engine, secure storage for keys, and a random number generator. For application development the fleck platform supports TinyOS 1.x.

8.2.8   Cricket

Cricket is a sensor-based device used for indoor localization. It was developed by MIT scientists [17] and manufactured by MEMSIC [26]. It has been widely utilized in research conducted in indoor environments where distance estimation between sensor nodes is needed. There are several advantages of the system including small device size, high measurement accuracy, scalability, user privacy and ease of deployment [26]. A cricket node can be configured as a beacon or a listener. The most common method to use Cricket is to attach the beacons in the infrastructure (walls and/or ceilings) and set listeners on a mobile device that is going to be localized or navigated.

This sensor-based system can provide distance estimates using time-difference-of-arrival (TDoA) with an accuracy of within 1–3 cm. The system utilizes the combination of RF and ultrasound signal to measure distance and provide location information.

The Cricket mote is built on a MICA2 low-power consumption sensor board combined with an RF module and associated hardware to implement the distance measurement feature and the ability for sending distance information. Figures 8.7 and 8.8 show the Cricket hardware implementation graph and the real Cricket mote.

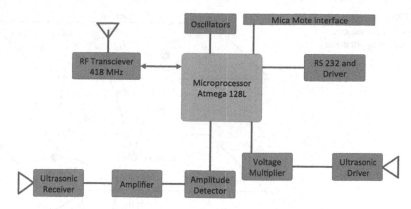

Fig. 8.7 Cricket hardware implementation block diagram with its main components [13]

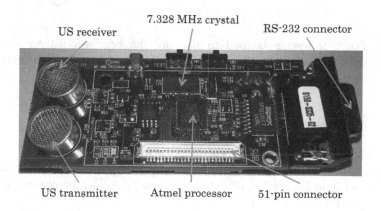

Fig. 8.8 Cricket mote [13]

Cricket mote uses ATmega128L as a main processor, which belongs to the AVR series. It is a high-performance, low-power 8-bit microcontroller that runs at 7.37 MHz in active mode as well as a 32.768 kHz clock that is applied in a power saving mode. The mote also has 128 KB programmable Flash ROM, 4 KB read-only EEPROM, and 53 general-purpose lines. The Cricket uses CC1000 as a low-power RF transceiver, which operates in the frequency range of 300–1000 MHz.

The Cricket system has an ultrasound unit (receiver and transmitter). Within the transmitter, there is a transducer to generate piezoelectric ultrasonic pulses at 40 kHz with configurable duration.

In a typical application using the Cricket system, there are multiple beacons deployed on walls or ceilings that periodically transmit RF and ultrasound signal, Fig. 8.9. At least one listener attached to the host device (laptop, robot) passively receives those signals to calculate the distance between each beacon and itself.

Fig. 8.9 Deployed cricket beacons and listeners in a localization application

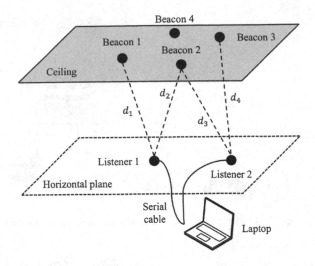

The TDoA technique is used to measure the beacon-to-listener distances. The listener estimates distance by obtaining the TDoA of the RF and the ultrasound signals that is given by:

$$\Delta T = \frac{d}{v_{US}} - \frac{d}{v_{RF}}, \tag{8.1}$$

where d is the distance from the beacon to the listener, v_{US} is the speed of sound propagation (in a normal room temperature and humidity $v_{US} \cong 344$ m/s) and v_{RF} is the speed of light ($v_{RF} \cong 3 \times 10^8$ m/s). With $v_{RF} > > v_{US}$, the distance is given by an approximation: $d \approx \Delta T \times v_{US}$.

8.2.9 Shimmer Wireless Node

Developed by Shimmer Research™, the Shimmer Wireless Platform was designed from the ground up to be a low-power, modular, wearable sensor node for use in healthcare, sports science, environmental sensing, and biofeedback [25]. A typical platform would include the main unit and a sensor board. The packaging is designed so that the main board rests at the bottom of the package and the sensor board rests on the top.

The main node consists of a Texas Instruments MSP430F1611 microcontroller featuring an 8 MHz clock, 2 DAC outputs, and an 8-channel, 12-bit ADC. The microcontroller also includes 10 KB of RAM and 48 KB of flash memory. Two sensors are integrated onto the main node (three-axis accelerometer and a tilt/vibration switch). The Shimmer Wireless Platform uses dual radios including TI CC2420 radio chip for 2.4 GHz IEEE 802.15.4 communication and the Bluetooth

Fig. 8.10 XM 1000 wireless sensor node (reproduced by permission of ADVANTICSYS, [32])

radio. Power is provided to the node with a Li-ion rechargeable battery. The node will accept voltages from 2.2 to 3.6 V.

Some examples of sensor modules include the 9DoF kinematic, GPS, and ECG. The 9DoF kinematic sensor board features a Honeywell HMC5843 three-axis digital compass and an InvenSense 500 series MEMS-based gyro. These sensors allow for complex motion sensing in a 3D environment. Worn on a user, it enables gestural computing and can be used for virtual reality and gaming applications.

8.2.10 ADVANTICSYS XM1000

The ADVANTICSYS XM1000 is a wireless sensor node based on an upgraded Crossbow TelosB platform [32] and is powered by two AA batteries, Fig. 8.10. It features the 16 MHz, 16-bit TI MSP430F2618 microcontroller using the RISC instruction set. Integrated into the microcontroller are 116 KB of program flash and 8 KB of data RAM with node-integrated external flash chip. The expanded memory allows for over-the-air reprogramming of the device as well as implementation of Device Profile for Web Services (DPWS)—a method for web service discovery and communication with the device on a LAN or WAN. The microcontroller enables communication using UART, SPI, and I^{2}C protocols.

Several sensors are directly integrated onto the board for monitoring temperature, humidity, and light. Wireless connectivity at 2.4 GHz is enabled by the TI CC2420 transceiver. This chip is IEEE 802.15.4 compliant, and has a range of up to 120 m outdoors and 30 m indoors.

8.3 WSN Simulation Tools

When conducting research on WSNs, it is often more feasible to use WSN simulations than deploy a real WSN. One common reason for preferring a WSN simulation is that of minimizing costs—for example, the total cost of purchasing WSN nodes and other hardware would be prohibitive for applications requiring very large number (e.g., thousands) of nodes. In such cases, researchers use WSN simulations, with the decision to deploy the real nodes depending on the promise depicted by the simulation-based research. Cost limitations are however not the only reason in support of WSN simulations. Other reasons include [5]: (1) *Network Debugging Issues*—debugging large distributed networks can sometimes be a daunting task and simulation can, for certain scenarios, provide a means to find and correct bugs prior to undertaking real WSN implementations. (2) *Harsh Operating Environments*—the target environments for certain WSN applications are unsafe for humans. Examples of applications associated with such environments include those for monitoring wildlife, volcanic activity or adverse weather conditions to mention but a few. In such cases, simulation offers a means to conduct experiments in a safe environment before resources and necessary mitigations against threats can be dedicated to deployments in the real operation environment. (3) *The Need for Precise Control of WSN Parameters*—in WSN research, it is often required to evaluate the behavior of the network for different combinations of precisely chosen parameter settings. Because a live WSN setting is not entirely controllable by the experimenter, simulations come in handy to provide this controlled and repeatable environment.

To address challenges such as these, the community has put forth a number of simulators. This section briefly explores five of the most popular WSN simulation tools.

8.3.1   ns-2 (Network Simulator-2)

Network Simulator-2 (ns-2) is an open-source discrete-event simulator. The simulation kernel, models and protocols are implemented in C++ while the creation, control and management of simulations is done in Object-oriented Tool Command Language (OTcl) [5, 8]. Network Simulator-2 was designed for simulating traditional IP networks and as such requires special extensions in order to support WSN simulations. The extensions augment the base ns-2 functionality with features such as sensing, processing energy consumption, WSN operation modes (e.g., sleep and wake-up modes), variations in node capabilities (e.g., regular nodes versus access points), and various options for dissemination of sensed data among others. Two of these extensions such as MannaSim [14] and SensorSim [33] have been widely used in past WSN research. SensorSim has however long ceased to be developed or supported and we do not discuss it here.

MannaSim comprises two components: (1) the MannaSim Framework, which encompasses the core extension module used for the design, development and analysis of different WSN applications; and (2) the Script Generator tool, which provides a front-end via which Tool Command Language (Tcl) simulation scripts are easily created. MannaSim enables the user to control aspects of the network's composition (e.g., number, type and density of nodes) and its organization (e.g., flat or hierarchical network). It supports a wide range of applications and provides a testbed for various algorithms and protocols [14]. MannaSim comprises a set of classes, which extend the corresponding ns-2 classes. For example, the *Battery* class extends ns-2's *EnergyModel* class and provides methods for the implementation of various battery models. Another example is the *SensorNode* class, which extends ns-2's *MobileNode* class by adding features such as sensing and processing. A full list of classes can be found in [14, 20].

One of the major advantages of ns-2 as a WSN simulator is its abundance of publicly available protocols and algorithms [22]. Some of its most notable weaknesses on the other hand include the steep learning curve that one typically goes through before undertaking meaningful simulations and the absence of an application model [22].

8.3.2 OMNETT++

OMNETT++ is a discrete-event simulation environment for communication networks in general. Its components are developed in C++ while the simulation implementation is based on a high level language called Network Description Language (NED) [20]. Because OMNETT++ is not specifically designed for WSNs, it requires special packages to be able to run WSN simulations. Below, we briefly discuss two of the most popular amongst these frameworks.

Castalia: Via a wide range of tunable parameters, Castalia can be used to simulate a broad spectrum of WSN platforms. Its most outstanding features include [34]: (i) the advanced channel model which is based on empirically measured data; (ii) the radio model which is based on real radio components; (iii) its extended modeling sensing provisions, and (iv) its intrinsic design for adaptability and extensibility. The latter attribute in particular enables researchers to easily port their algorithms and protocols into Castalia. Castalia is not recommended for cases where emphasis of simulation is to observe detailed platform-specific behavior. In such cases, Castalia is best used as a first-line simulator that provides a coarse-grained view of the WSN's behavior before more fine-grained platform-specific simulations can be run [34].

MiXiM (Mixed Simulator): It is a merger of several OMNETT++ based frameworks for mobile and wireless simulation [35]. The word "mixed" in its name comes from its being a combination of various simulators [20]. Its radio propagation model is based on the Channel Simulator (ChSim) [36], connection management and mobility support are based on the Mobility Framework (MF) [37] while the protocol library is derived from the MAC simulator [38], the Positif *Framework* [38] and from the *Mobility Framework*. MiXiM supports models based on both 2D and 3D settings and, in addition to traditional WSN nodes, supports the simulation of objects such as walls and houses as obstacles to the propagation of radio waves.

8.3.3 TinyOS Simulator (TOSSIM)

TOSSIM is a discrete-event simulator specifically designed for TinyOS applications. It has support for two programming interfaces, one of which is in Python, the other in C++. Using the Python interface, one can interact with the running simulation dynamically. TOSSIM programs can replace entire components of TinyOS applications with their simulation implementations since these programs in general require no modification to be run on the motes [1, 28]. This feature gives developers a big margin for application testing and debugging since TinyOS applications may be developed and compiled to the TOSSIM framework running on a desktop. TOSSIM captures details of TinyOS's behavior, closely simulating each ADC capture and each interrupt to the system. Using tools external to TOSSIM, users can implement models of different real-world phenomena.

The design choice to keep real-world models external to the simulator was aimed to allow the flexibility for researchers to implement their own models in such a way that the TOSSIM environment imposes no definitions of its own regarding what is correct or wrong [11]. Despite capturing TinyOS behavior at a very fine level, it is noteworthy that TOSSIM makes several simplifying assumptions that could cause unexpected behavior. For example, as a direct consequence of being a discrete-event simulator, TOSSIM interrupts are non pre-emptive. If, as an example, pre-emption were to put a real-world mote into unrecoverable state, an equivalent simulated TOSSIM mote would not capture this behavior [11]. These kinds of challenges notwithstanding, TOSSIM offers a good first step towards the understanding of algorithm performance prior to implementation on a real WSN.

8.3.4 Optimized Network Engineering Tool (OPNET)

OPNET is a proprietary[1] discrete-event simulator targeted for computer networks in general. Different from simulators such as ns-2, OPNET supports the modeling of sensor hardware (e.g., transceivers and antennas), and has provision for the definition of custom packet formats [1]. Using its GUI, one may model, graph or animate the simulator's output. Researchers have designed an interface for compiling TinyOS applications to OPNET models [21]. Just as the kind of interoperability realizable between TOSSIM and TinyOS, this interface enables a shared code model in which the same application code is shared between the TinyOS executable and the OPNET simulation model. The interface has support for scenario management and statistics management, has the ability for instantiations of different applications to be simulated together in the same memory space, and has the availability of a much larger wealth of models, the combination of which give OPNET an edge when it comes to sophisticated TinyOS simulations [21].

---

[1]Since October 2012, the simulator is owned by Riverbed [40].

Typical of proprietary products, the OPNET license comes with a considerable amount of documentation and study cases that can be useful to researchers [1].

8.3.5 Avrora

Avrora is a Java-based open-source simulator for embedded sensing programs. It simulates the actual microcontroller programs (as opposed to models of the software), and executes cycle-accurate simulations of the devices and the radio communication [23]. It can scale to networks of up to 10,000 nodes and can handle as many as 25 nodes in real-time [23]. It has a nearly complete implementation of the mica2 hardware platform, an ATMega128L implementation, and an implementation of the CC1000 AM radio [39]. To enable testing, debugging, or analyzing of programs before running them in network simulations, Avrora supports the simulation of a sensor network program on a single node. For complete network simulation Avrora provides full timing accuracy and allows programs to communicate via the CC1000 AM radio. Two notable challenges seen with Avrora are that it is slow (e.g., it is 50 % slower than TOSSIM) and it does not model node mobility or clock drift [22].

References

1. A. Abuarqoub, F. Al-Fayez, T. Alsboui, M. Hammoudeh, and A. Nisbe, "Simulation issues in wireless sensor networks: a survey," *Proc. the Sixth International Conference on Sensor Technologies and Applications (SENSORCOMM 2012)*, Rome, Italy, 2012.
2. P. Corke, T. Wark, R. Jurdak, W. Hu, P. Valencia, and D. Moore, "Environmental wireless sensor networks," *Proceedings of the IEEE, Special Issue on Emerging Sensor Network Applications*, vol. 98, no. 11, pp. 1903–1917, November 2010.
3. *Cricket User Manual*, MIT Computer Science and Artificial Intelligence Lab, Second Edition, Jan. 2005.
4. C.-M. Hsieh, F. Samie, M.S. Srouji, M. Wang, Z. Wang, and J. Henkel, "Hardware/software co-design for a wireless sensor network platform," *2014 International Conference on Hardware/Software Codesign and System Synthesis (CODES + ISSS)*, New Delhi, 2014.
5. M. Jevtic, N. Zogovic, and G. Dimic, Evaluation of Wireless Sensor Network Simulators, *17th Telecommunications forum TELFOR*, Belgrade, Serbia, 2009.
6. M. Johnson, M. Healy, P. van de Ven, M.J. Hayes, J. Nelson, T. Newe, and E. Lewis, "A comparative review of wireless sensor network mote technologies," *Proc. IEEE Sensors*, pp. 1439–1442, October 2009.
7. S.D.T. Kelly, N.K. Suryadevara, and S.C. Mukhopadhyay, "Towards the Implementation of IoT for Environment Condition Monitoring in Homes," *IEEE Sensors Journal*, vol. 13, no. 10, October 2013.
8. S. Khan, A.K. Pathan, and N.A. Alrajeh (Editors), *Wireless Sensor Networks: Current Status and Future Trends*, CRC Press, Taylor & Francis Group, Boca Raton, Florida, 2013.
9. J. Ko, K. Klues, C. Richter, W. Hofer, B. Kusy, M. Bruenig, T. Schmid, Q. Wang, P. Dutta and A. Terzis, "Low power or high performance? A tradeoff whose time has come (and nearly gone)," *Proc. the 9th European Conference on Wireless Sensor Networks*, Italy 2012.

10. M. Kohvakka, T. Arpinen, M. Haunikainen and T.D. Hamalainen, "High-performance multi-radio wsn platform," *Proceedings of the 2nd International Workshop on Multi-hop Ad Hoc Networks: From Theory to Reality*, pp. 95–97, 2006.

11. P. Levis, N. Lee, M. Welsh, and D. Culler, "TOSSIM: accurate and scalable simulation of entire TinyOS applications," *Proc. of the 1st International Conference on Embedded Networked Sensor Systems*, New York, NY, USA, 2003.

12. A. Mainwaring, J. Polastre, R. Szewczyk, D. Culler, and J. Anderson, "Wireless Sensor Networks for Habitat Monitoring," *Proc. the 1st ACM International Workshop on Wireless Sensor Networks and Applications (WSNA '02)*, pp. 88–97, 2002.

13. MIT Computer Science and Artificial Intelligence Lab, "Cricket user Manual, Second Edition," January 2005.

14. R.M. Pereira, L.B. Ruiz, and M.L.A. Ghizoni, "MannaSim: A NS-2 Extension to Simulate Wireless Sensor Networks," The Fourteenth International Conference on Networks, Barcelona, Spain, 2015.

15. F. Philipp, F. A. Samman and M. Glesner, "Design of an autonomous platform for distributed sensing-actuating systems", *22nd IEEE International Symposium on Rapid System Prototyping (RSP)*, pp. 85–90, 2011.

16. J. Portilla, T. Riesgo and A. De Castro, "A reconfigurable FPGA-based architecture for modular nodes in wireless sensor net-works," *3rd Southern Conference on Programmable Logic*, Mar del Plata, Argentina, 2007.

17. N.B. Priyantha, *The Cricket Indoor Location System*, Ph.D. Dissertation, Massachusetts Institute of Technology, Jun. 2005.

18. P. Sikka, P. Corke, L. Overs, P. Valencia, and T. Wark, "Fleck—a platform for real-world outdoor sensor networks," *Proc. 3rd International Conference on Intelligent Sensors, Sensor Networks and Information*, Melbourne, Australia, pp. 709–714, December 2007.

19. G. Stamatescu, C. Chiṭu, C. Vasile, I. Stamatescu, D. Popescu and V. Sgârciu, "Analytical and experimental sensor node energy modeling in ambient monitoring," 9th *IEEE Conference on Industrial Electronics and Applications (ICIEA)*, Hangzhou, China, 2014.

20. M. Stehlik, *Comparison of Simulators for Wireless Sensor Networks*, Master Thesis, Masaryk University, 2011.

21. D. Sumorok, D. Starobinski, and A. Trachtenberg, "Simulation of TinyOS Wireless Sensor Networks Using OPNET," *Proc. of OPNETWORK 04*, Washington DC, August 2004.

22. H. Sundani, H. Li, V.K. Devabhaktuni, M. Alam, and P. Bhattacharya, "Wireless sensor network simulators, a survey and comparisons," *International Journal of Computer Networks (IJCN)*, vol. 2, no. 5, 2015.

23. B.L. Titzer, D.K. Lee, and J. Palsberg, "Avrora: scalable sensor network simulation with precise timing," *Proc. the Fourth International Symposium on Information Processing in Sensor Networks*, UCLA, Los Angeles, CA, 2005.

24. J.S.C. Turner, M.F. Ramli, L.M. Kamarudin, A. Zakaria, A.Y.M. Shakaff, D.L. Ndzi, C.M. Nor, N. Hassan, and S.M Mamduh, "The study of human movement effect on signal strength for indoor WSN deployment," *IEEE Conference on Wireless Sensors (ICWiSe2013)*, Kuching, Sarawak, December 2013.

25. "Shimmer Wireless Sensor Unit/Platform." [Online]. Available: http://www.shimmer-research.com/p/products/sensor-units-and-modules/shimmer-wireless-sensor-unitplatform.

26. http://www.memsic.com/wireless-sensor-networks/.

27. http://www.digi.com/lp/xbee.

28. http://www.tinyos.net/.

29. http://wsn.cse.wustl.edu/images/e/e3/Imote2_Datasheet.pdf.

30. http://www.eistec.se/mulle/.

31. http://www.coalesenses.com.

32. http://www.advanticsys.com.

33. http://www.nrl.navy.mil/itd/ncs/products/sensorsim.

34. https://castalia.forge.nicta.com.au/index.php/en/.

35. http://mixim.sourceforge.net/.

36. http://www-old.cs.uni-paderborn.de/en/fachgebiete/research-group-computer-networks/projects/chsim.html.
37. http://mobility-fw.sourceforge.net/.
38. http://www.consensus.tudelft.nl/software.html.
39. http://compilers.cs.ucla.edu/avrora/sensors.html.
40. http://www.riverbed.com.